# SpringerBriefs in Electrical and Computer Engineering

T0238968

For further volumes:
http://www.springer.com/series/10059

S Krishna

# An Introduction to Modelling of Power System Components

Springer

S Krishna
Department of Electrical Engineering
Indian Institute of Technology Madras
Chennai, Tamil Nadu
India

ISSN 2191-8112          ISSN 2191-8120   (electronic)
ISBN 978-81-322-1846-3   ISBN 978-81-322-1847-0   (eBook)
DOI 10.1007/978-81-322-1847-0
Springer New Delhi Heidelberg New York Dordrecht London

Library of Congress Control Number: 2014933281

Printed on acid-free paper

Springer is part of Springer Science+Business Media (www.springer.com)

# Preface

The book is based on the notes prepared for the courses taught by the author at Indian Institute of Technology Madras. The book gives the derivation of the model of power system components such as synchronous generator, transformer, transmission line, DC transmission system, flexible AC transmission systems, excitation system, and speed governor. The model of load and prime movers are given without derivation. The book can serve as a text for a short graduate course on power system modelling, or as a supplement for graduate courses on power system stability and flexible AC transmission systems.

Chennai, India, December 2013                                      S Krishna

Preface

# Contents

# Chapter 1
# Synchronous Generator

## 1.1 Construction

The synchronous generator converts mechanical energy into electrical energy. It has a stationary component or stator, and a rotating component or rotor. The stator is an annular structure made up of iron and has slots. Insulated coils are placed in the slots, and these coils are connected to obtain a three-phase winding. The rotor has electromagnets which are known as field poles. The rotor is placed within the stator. The cross section of a synchronous generator with two field poles is shown in Fig. 1.1. $a$, $b$, and $c$ are the three stator windings which are 120° apart. $f$ is the field winding. A dot indicates that positive current flow is out of the paper/screen, while a cross indicates that positive current flow is into the paper/screen. If the field winding is excited by a DC source, then as the rotor rotates, an emf is induced in the armature winding according to Faraday's law.

The number of field poles depends on the speed of the prime mover. If the number of field poles is $p_f$, then one rotation of the rotor induces $p_f/2$ cycles of emf in the armature winding. The hydraulic turbines operate at low speed. Therefore, to obtain the rated frequency, the synchronous generator driven by a hydraulic turbine has a large number of field poles. On the other hand, steam and gas turbines operate at high speeds; hence, the synchronous generator driven by these turbines has two or four field poles.

The rotors often have amortisseur or damper circuits in the form of copper or brass rods. These rods are short-circuited and are intended to damp out oscillations in speed.

## 1.2 Model with Three Damper Windings

The model is initially developed for a synchronous generator having a pair of field poles; generalization to any number of field poles is done later. An axis is defined for each stator winding as shown in Fig. 1.2. Two axes, namely direct axis ($d$-axis)

An erratum to this chapter is available at DOI 10.1007/978-81-322-1847-0_5

S Krishna, *An Introduction to Modelling of Power System Components*,
SpringerBriefs in Electrical and Computer Engineering,
DOI: 10.1007/978-81-322-1847-0_1, © The Author(s) 2014

**Fig. 1.1** Synchronous
generator

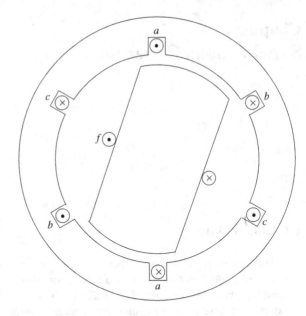

**Fig. 1.2** Axes of
a synchronous generator

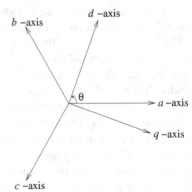

and quadrature axis ($q$-axis), are also defined. The $d$-axis is along the field pole. The $q$-axis lags the $d$-axis by 90° [1]. $a$-axis, $b$-axis, and $c$-axis are stationary, and $d$-axis and $q$-axis rotate at the speed of the rotor. The rotor is assumed to rotate in the counterclockwise direction, and $\theta$ is the angle by which the $d$-axis leads the $a$-axis.

The amortisseur circuits and the eddy current effects in the rotor are represented by two equivalent sets of short-circuited damper windings [1–3]. One set of windings is oriented such that the flux in the rotor due to current in these windings is along $d$-axis; these windings are said to be on the $d$-axis. The other set of windings is oriented such that the flux in the rotor due to current in these windings is along $q$-axis; these windings are said to be on the $q$-axis. The damper winding $1d$ is on the $d$-axis, and the damper windings $1q$ and $2q$ are on the $q$-axis.

**Fig. 1.3** Circuit diagram of a winding

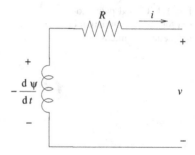

## 1.2.1 Voltage, Current, and Flux Linkage Relationships

The circuit diagram of a synchronous generator winding ($a, b, c, f, 1d, 1q$, or $2q$) is shown in Fig. 1.3 [1]. $\psi$, $R$, $i$, and $v$ are flux linkage, resistance, current, and voltage, respectively.

By Kirchhoff's voltage law,

$$\frac{d\psi_a}{dt} = -R_a i_a - v_a \tag{1.1}$$

$$\frac{d\psi_b}{dt} = -R_a i_b - v_b \tag{1.2}$$

$$\frac{d\psi_c}{dt} = -R_a i_c - v_c \tag{1.3}$$

$$\frac{d\psi_f}{dt} = -R_f i_f + v_f \tag{1.4}$$

$$\frac{d\psi_{1d}}{dt} = -R_{1d} i_{1d} \tag{1.5}$$

$$\frac{d\psi_{1q}}{dt} = -R_{1q} i_{1q} \tag{1.6}$$

$$\frac{d\psi_{2q}}{dt} = -R_{2q} i_{2q} \tag{1.7}$$

The polarity of $v_f$ is such that $v_f$ and $i_f$ are positive in steady state.

The flux linkages, currents, and inductances are related as follows:

$$
\begin{bmatrix} \psi_a \\ \psi_b \\ \psi_c \\ \psi_f \\ \psi_{1d} \\ \psi_{1q} \\ \psi_{2q} \end{bmatrix}
=
\begin{bmatrix}
L_a & M_{ab} & M_{ac} & M_{af} & M_{a1d} & M_{a1q} & M_{a2q} \\
M_{ba} & L_b & M_{bc} & M_{bf} & M_{b1d} & M_{b1q} & M_{b2q} \\
M_{ca} & M_{cb} & L_c & M_{cf} & M_{c1d} & M_{c1q} & M_{c2q} \\
M_{fa} & M_{fb} & M_{fc} & L_f & M_{f1d} & M_{f1q} & M_{f2q} \\
M_{1da} & M_{1db} & M_{1dc} & M_{1df} & L_{1d} & M_{1d1q} & M_{1d2q} \\
M_{1qa} & M_{1qb} & M_{1qc} & M_{1qf} & M_{1q1d} & L_{1q} & M_{1q2q} \\
M_{2qa} & M_{2qb} & M_{2qc} & M_{2qf} & M_{2q1d} & M_{2q1q} & L_{2q}
\end{bmatrix}
\begin{bmatrix} i_a \\ i_b \\ i_c \\ i_f \\ i_{1d} \\ i_{1q} \\ i_{2q} \end{bmatrix}
\tag{1.8}
$$

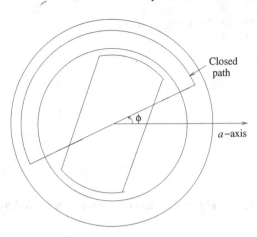

Closed
path

$a$−axis

$\phi$

## 1.2.2 Expression for Inductances

The stator windings are assumed to be filamentary. Let each stator winding have $N$
turns. The permeability of the stator core and the rotor core is assumed to be infinite.
Therefore, the magnetic field intensity is nonzero only in the air gap. If $\phi$ is the angle
measured from the $a$-axis in the counterclockwise direction, the air gap length $g$ is
a periodic function of $\phi$ which satisfies the condition

$$g(\phi + \pi) = g(\phi) \tag{1.9}$$

$g$ is small compared to the inner radius of the stator. Therefore, it is assumed that
the magnetic field intensity in the air gap is radial [4]. From (1.9), it follows that the
magnetic field intensity $H_a$ due to $i_a$, in the air gap, in the radial outward direction,
satisfies the following equation:

$$H_a(\phi + \pi) = -H_a(\phi) \tag{1.10}$$

The expression for $H_a$ is obtained from Ampere's law applied to the closed path
shown in Fig. 1.4. The closed path consists of a semicircle in the stator core and a
straight line passing through and perpendicular to the rotor axis of rotation.
Therefore,

$$H_a = \begin{cases} \frac{Ni_a}{2g} & \text{if } -\frac{\pi}{2} < \phi < \frac{\pi}{2} \\ -\frac{Ni_a}{2g} & \text{if } \frac{\pi}{2} < \phi < \frac{3\pi}{2} \end{cases} \tag{1.11}$$

A quantity called air gap magnetomotive force (MMF) is defined as the product of
air gap magnetic field intensity and air gap length. The air gap MMF due to $i_a$ is

$$F_a \triangleq H_a g \tag{1.12}$$

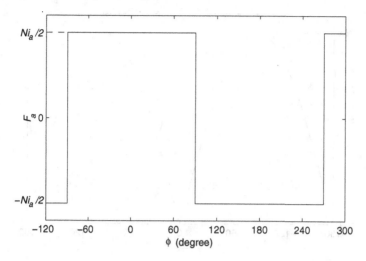

**Fig. 1.5**  Air gap MMF due to $i_a$

The variation in $F_a$ with $\phi$ is shown in Fig. 1.5.

Let the air gap MMF be approximated by its fundamental component with peak value $2Ni_a/\pi$ [4]. This approximation can be done even for a practical stator winding placed in the stator core having many slots and not restricted to six slots as in Fig. 1.1. In fact, the air gap MMF for a practical stator winding is closer to a sinusoidal waveform than that for the one shown in Fig. 1.1. In order to establish a sinusoidal air gap MMF waveform, it will be shown that the equivalent winding must be sinusoidally distributed. Suppose the number of turns per radian at any location is

$$n_a = \begin{cases} -\frac{2}{\pi}N\sin\phi & \text{if } -\pi \le \phi \le 0 \\ \frac{2}{\pi}N\sin\phi & \text{if } 0 \le \phi \le \pi \end{cases} \tag{1.13}$$

The winding distribution is shown in Fig. 1.6. The current $i_a$ in the turns for $-\pi < \phi < 0$ is into the paper/screen, whereas the current in the turns for $0 < \phi < \pi$ is out of the paper/screen. The number of turns of the equivalent sinusoidally distributed winding is

$$N_a = \int_0^\pi n_a d\phi = \int_0^\pi \frac{2}{\pi}N\sin\phi d\phi = \frac{4}{\pi}N \tag{1.14}$$

Substituting this in (1.13) gives

$$n_a = \begin{cases} -\frac{N_a}{2}\sin\phi & \text{if } -\pi \le \phi \le 0 \\ \frac{N_a}{2}\sin\phi & \text{if } 0 \le \phi \le \pi \end{cases} \tag{1.15}$$

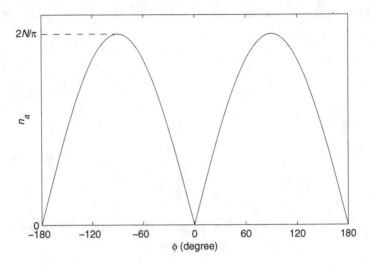

**Fig. 1.6** Winding distribution for sinusoidal air gap MMF

The air gap magnetic field intensity $H_{a1}$ in the radial outward direction, due to $i_a$ with a winding distribution given by (1.15), is obtained from Ampere's law applied to the closed path shown in Fig. 1.4. For $-\pi \leq \phi \leq 0$,

$$
\begin{aligned}
H_{a1} &= \frac{1}{2g}\left(-i_a \int_{\phi}^{0} n_a \mathrm{d}\beta + i_a \int_{0}^{\phi+\pi} n_a \mathrm{d}\beta\right) \\
&= \frac{1}{2g}\left[-i_a \int_{\phi}^{0}\left(-\frac{N_a}{2}\right)\sin\beta\mathrm{d}\beta + i_a \int_{0}^{\phi+\pi}\frac{N_a}{2}\sin\beta\mathrm{d}\beta\right] \\
&= \frac{N_a i_a}{2g}\cos\phi
\end{aligned}
\tag{1.16}
$$

Since $H_{a1}(\phi+\pi) = -H_{a1}(\phi)$, the expression for $H_{a1}$ given by (1.16) is applicable for any $\phi$. The air gap MMF due to $i_a$ with the winding distribution given by (1.15) is

$$
F_{a1} = H_{a1}g = \frac{N_a i_a}{2}\cos\phi
\tag{1.17}
$$

The variation in $F_{a1}$ with $\phi$ is shown in Fig. 1.7. Hence, the winding distribution given by (1.15) results in the air gap MMF equal to the fundamental component of $F_a$.

The distribution of the other equivalent windings ($b, c, f, 1d, 1q$, and $2q$) and the air gap MMFs are given by

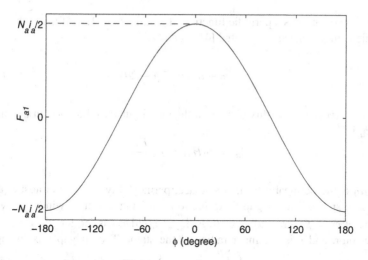

**Fig. 1.7** Fundamental component of air gap MMF due to $i_a$

$$n_b = \frac{N_a}{2} \sin\left(\phi - \frac{2\pi}{3}\right) \quad \text{if } \frac{2\pi}{3} \leq \phi \leq \frac{5\pi}{3} \tag{1.18}$$

$$n_c = \frac{N_a}{2} \sin\left(\phi + \frac{2\pi}{3}\right) \quad \text{if } -\frac{2\pi}{3} \leq \phi \leq \frac{\pi}{3} \tag{1.19}$$

$$n_f = \frac{N_f}{2} \sin(\phi - \theta) \quad \text{if } \theta \leq \phi \leq \theta + \pi \tag{1.20}$$

$$n_{1d} = \frac{N_{1d}}{2} \sin(\phi - \theta) \quad \text{if } \theta \leq \phi \leq \theta + \pi \tag{1.21}$$

$$n_{1q} = \frac{N_{1q}}{2} \cos(\phi - \theta) \quad \text{if } \theta - \frac{\pi}{2} \leq \phi \leq \theta + \frac{\pi}{2} \tag{1.22}$$

$$n_{2q} = \frac{N_{2q}}{2} \cos(\phi - \theta) \quad \text{if } \theta - \frac{\pi}{2} \leq \phi \leq \theta + \frac{\pi}{2} \tag{1.23}$$

$$F_{b1} = \frac{N_a}{2} i_b \cos\left(\phi - \frac{2\pi}{3}\right) \tag{1.24}$$

$$F_{c1} = \frac{N_a}{2} i_c \cos\left(\phi + \frac{2\pi}{3}\right) \tag{1.25}$$

$$F_{f1} = \frac{N_f}{2} i_f \cos(\phi - \theta) \tag{1.26}$$

$$F_{1d1} = \frac{N_{1d}}{2} i_{1d} \cos(\phi - \theta) \tag{1.27}$$

$$F_{1q1} = -\frac{N_{1q}}{2} i_{1q} \sin(\phi - \theta) \tag{1.28}$$

$$F_{2q1} = -\frac{N_{2q}}{2} i_{2q} \sin(\phi - \theta) \tag{1.29}$$

The air gap length is a periodic function of $\phi$ satisfying (1.9). If the higher-order harmonic components are neglected [4], then

$$\frac{1}{g} = a_0 + a_2 \cos(2\phi - 2\theta) \tag{1.30}$$

where $a_0 > a_2 > 0$. The flux density in the radial outward direction in the air gap due to $i_a$ is

$$B_a = \mu_0 H_{a1} = \mu_0 \frac{F_{a1}}{g} \tag{1.31}$$

where $\mu_0$ is the permeability of free space; permeability of air is almost equal to that of free space. The air gap flux, due to $i_a$, linking a turn of winding $a$ whose sides are at $\phi = \beta - \pi$ and $\phi = \beta$, is $\int_{\beta-\pi}^{\beta} B_a r l d\phi$; $l$ is the length of the stator and the rotor, and $r$ is the inner radius of the stator. The air gap flux linkage of winding $a$, due to $i_a$, is $\int_0^{\pi} n_a \left( \int_{\beta-\pi}^{\beta} B_a r l d\phi \right) d\beta$. The ratio of this flux linkage to $i_a$ gives the self inductance of winding $a$ due to the flux crossing the air gap. The total self inductance of winding $a$ is obtained by adding to this, the leakage inductance due to the leakage flux. If the leakage inductance of winding $a$ is $L_{al}$, the total self inductance of winding $a$ is

$$L_a = L_{al} + \int_0^{\pi} \frac{N_a}{2} \sin \beta \left[ \int_{\beta-\pi}^{\beta} \mu_0 \frac{N_a}{2} \cos \phi \left\{ a_0 + a_2 \cos (2\phi - 2\theta) \right\} r l \, d\phi \right] d\beta$$

$$= L_{a0} + L_{a2} \cos (2\theta) \tag{1.32}$$

where

$$L_{a0} \triangleq L_{al} + \frac{N_a^2}{4} \pi \mu_0 r l a_0, \quad L_{a2} \triangleq \frac{N_a^2}{8} \pi \mu_0 r l a_2 \tag{1.33}$$

The expressions for other inductances are obtained as follows:

$$M_{ab} = \int_0^{\pi} \frac{N_a}{2} \sin \beta \left[ \int_{\beta-\pi}^{\beta} \mu_0 \frac{N_a}{2} \cos \left( \phi - \frac{2\pi}{3} \right) \left\{ a_0 + a_2 \cos (2\phi - 2\theta) \right\} r l \, d\phi \right] d\beta$$

$$= M_{ab0} + L_{a2} \cos \left( 2\theta - \frac{2\pi}{3} \right) \tag{1.34}$$

$$M_{ac} = \int_0^{\pi} \frac{N_a}{2} \sin \beta \left[ \int_{\beta-\pi}^{\beta} \mu_0 \frac{N_a}{2} \cos \left( \phi + \frac{2\pi}{3} \right) \left\{ a_0 + a_2 \cos (2\phi - 2\theta) \right\} r l \, d\phi \right] d\beta$$

$$= M_{ab0} + L_{a2} \cos\left(2\theta + \frac{2\pi}{3}\right) \tag{1.35}$$

$$M_{af} = \int_0^\pi \frac{N_a}{2} \sin\beta \left[ \int_{\beta-\pi}^\beta \mu_0 \frac{N_f}{2} \cos(\phi - \theta)\{a_0 + a_2 \cos(2\phi - 2\theta)\}\, rl\, d\phi \right] d\beta$$

$$= M_{af1} \cos\theta \tag{1.36}$$

$$M_{a1d} = \int_0^\pi \frac{N_a}{2} \sin\beta \left[ \int_{\beta-\pi}^\beta \mu_0 \frac{N_{1d}}{2} \cos(\phi - \theta)\{a_0 + a_2 \cos(2\phi - 2\theta)\}\, rl\, d\phi \right] d\beta$$

$$= M_{a1d1} \cos\theta \tag{1.37}$$

$$M_{a1q} = -\int_0^\pi \frac{N_a}{2} \sin\beta \left[ \int_{\beta-\pi}^\beta \mu_0 \frac{N_{1q}}{2} \sin(\phi - \theta)\{a_0 + a_2 \cos(2\phi - 2\theta)\}\, rl\, d\phi \right] d\beta$$

$$= M_{a1q1} \sin\theta \tag{1.38}$$

$$M_{a2q} = -\int_0^\pi \frac{N_a}{2} \sin\beta \left[ \int_{\beta-\pi}^\beta \mu_0 \frac{N_{2q}}{2} \sin(\phi - \theta)\{a_0 + a_2 \cos(2\phi - 2\theta)\}\, rl\, d\phi \right] d\beta$$

$$= M_{a2q1} \sin\theta \tag{1.39}$$

$$M_{ba} = \int_{2\pi/3}^{5\pi/3} \frac{N_a}{2} \sin\left(\beta - \frac{2\pi}{3}\right)$$

$$\times \left[ \int_{\beta-\pi}^\beta \mu_0 \frac{N_a}{2} \cos\phi \{a_0 + a_2 \cos(2\phi - 2\theta)\}\, rl\, d\phi \right] d\beta$$

$$= M_{ab0} + L_{a2} \cos\left(2\theta - \frac{2\pi}{3}\right) \tag{1.40}$$

$$L_b = L_{al} + \int_{2\pi/3}^{5\pi/3} \frac{N_a}{2} \sin\left(\beta - \frac{2\pi}{3}\right)$$

$$\times \left[ \int_{\beta-\pi}^{\beta} \mu_0 \frac{N_a}{2} \cos\left(\phi - \frac{2\pi}{3}\right) \{a_0 + a_2 \cos(2\phi - 2\theta)\} rl \, d\phi \right] d\beta$$

$$= L_{a0} + L_{a2} \cos\left(2\theta + \frac{2\pi}{3}\right) \tag{1.41}$$

$$M_{bc} = \int_{2\pi/3}^{5\pi/3} \frac{N_a}{2} \sin\left(\beta - \frac{2\pi}{3}\right)$$

$$\times \left[ \int_{\beta-\pi}^{\beta} \mu_0 \frac{N_a}{2} \cos\left(\phi + \frac{2\pi}{3}\right) \{a_0 + a_2 \cos(2\phi - 2\theta)\} rl \, d\phi \right] d\beta$$

$$= M_{ab0} + L_{a2} \cos(2\theta) \tag{1.42}$$

$$M_{bf} = \int_{2\pi/3}^{5\pi/3} \frac{N_a}{2} \sin\left(\beta - \frac{2\pi}{3}\right)$$

$$\times \left[ \int_{\beta-\pi}^{\beta} \mu_0 \frac{N_f}{2} \cos(\phi - \theta) \{a_0 + a_2 \cos(2\phi - 2\theta)\} rl \, d\phi \right] d\beta$$

$$= M_{af1} \cos\left(\theta - \frac{2\pi}{3}\right) \tag{1.43}$$

$$M_{b1d} = \int_{2\pi/3}^{5\pi/3} \frac{N_a}{2} \sin\left(\beta - \frac{2\pi}{3}\right)$$

$$\times \left[ \int_{\beta-\pi}^{\beta} \mu_0 \frac{N_{1d}}{2} \cos(\phi - \theta) \{a_0 + a_2 \cos(2\phi - 2\theta)\} rl \, d\phi \right] d\beta$$

$$= M_{a1d1} \cos\left(\theta - \frac{2\pi}{3}\right) \tag{1.44}$$

$$M_{b1q} = - \int_{2\pi/3}^{5\pi/3} \frac{N_a}{2} \sin\left(\beta - \frac{2\pi}{3}\right)$$

$$\times \left[ \int_{\beta-\pi}^{\beta} \mu_0 \frac{N_{1q}}{2} \sin(\phi - \theta) \{a_0 + a_2 \cos(2\phi - 2\theta)\} rl \, d\phi \right] d\beta$$

$$= M_{a1q1} \sin\left(\theta - \frac{2\pi}{3}\right) \tag{1.45}$$

$$M_{b2q} = -\int_{2\pi/3}^{5\pi/3} \frac{N_a}{2} \sin\left(\beta - \frac{2\pi}{3}\right)$$

$$\times \left[ \int_{\beta-\pi}^{\beta} \mu_0 \frac{N_{2q}}{2} \sin(\phi - \theta) \{a_0 + a_2 \cos(2\phi - 2\theta)\} rl \, d\phi \right] d\beta$$

$$= M_{a2q1} \sin\left(\theta - \frac{2\pi}{3}\right) \tag{1.46}$$

$$M_{ca} = \int_{-2\pi/3}^{\pi/3} \frac{N_a}{2} \sin\left(\beta + \frac{2\pi}{3}\right)$$

$$\times \left[ \int_{\beta-\pi}^{\beta} \mu_0 \frac{N_a}{2} \cos\phi \{a_0 + a_2 \cos(2\phi - 2\theta)\} rl \, d\phi \right] d\beta$$

$$= M_{ab0} + L_{a2} \cos\left(2\theta + \frac{2\pi}{3}\right) \tag{1.47}$$

$$M_{cb} = \int_{-2\pi/3}^{\pi/3} \frac{N_a}{2} \sin\left(\beta + \frac{2\pi}{3}\right)$$

$$\times \left[ \int_{\beta-\pi}^{\beta} \mu_0 \frac{N_a}{2} \cos\left(\phi - \frac{2\pi}{3}\right) \{a_0 + a_2 \cos(2\phi - 2\theta)\} rl \, d\phi \right] d\beta$$

$$= M_{ab0} + L_{a2} \cos(2\theta) \tag{1.48}$$

$$L_c = L_{al} + \int\limits_{-2\pi/3}^{\pi/3} \frac{N_a}{2} \sin\left(\beta + \frac{2\pi}{3}\right)$$

$$\times \left[ \int\limits_{\beta-\pi}^{\beta} \mu_0 \frac{N_a}{2} \cos\left(\phi + \frac{2\pi}{3}\right) \{a_0 + a_2 \cos(2\phi - 2\theta)\} rl \, d\phi \right] d\beta$$

$$= L_{a0} + L_{a2} \cos\left(2\theta - \frac{2\pi}{3}\right) \tag{1.49}$$

$$M_{cf} = \int\limits_{-2\pi/3}^{\pi/3} \frac{N_a}{2} \sin\left(\beta + \frac{2\pi}{3}\right)$$

$$\times \left[ \int\limits_{\beta-\pi}^{\beta} \mu_0 \frac{N_f}{2} \cos(\phi - \theta) \{a_0 + a_2 \cos(2\phi - 2\theta)\} rl \, d\phi \right] d\beta$$

$$= M_{af1} \cos\left(\theta + \frac{2\pi}{3}\right) \tag{1.50}$$

$$M_{c1d} = \int\limits_{-2\pi/3}^{\pi/3} \frac{N_a}{2} \sin\left(\beta + \frac{2\pi}{3}\right)$$

$$\times \left[ \int\limits_{\beta-\pi}^{\beta} \mu_0 \frac{N_{1d}}{2} \cos(\phi - \theta) \{a_0 + a_2 \cos(2\phi - 2\theta)\} rl \, d\phi \right] d\beta$$

$$= M_{a1d1} \cos\left(\theta + \frac{2\pi}{3}\right) \tag{1.51}$$

$$M_{c1q} = - \int\limits_{-2\pi/3}^{\pi/3} \frac{N_a}{2} \sin\left(\beta + \frac{2\pi}{3}\right)$$

$$\times \left[ \int\limits_{\beta-\pi}^{\beta} \mu_0 \frac{N_{1q}}{2} \sin(\phi - \theta) \{a_0 + a_2 \cos(2\phi - 2\theta)\} rl \, d\phi \right] d\beta$$

$$= M_{a1q1} \sin\left(\theta + \frac{2\pi}{3}\right) \tag{1.52}$$

$$M_{c2q} = - \int_{-2\pi/3}^{\pi/3} \frac{N_a}{2} \sin\left(\beta + \frac{2\pi}{3}\right)$$

$$\times \left[ \int_{\beta-\pi}^{\beta} \mu_0 \frac{N_{2q}}{2} \sin(\phi - \theta)\{a_0 + a_2 \cos(2\phi - 2\theta)\} rl\, d\phi \right] d\beta$$

$$= M_{a2q1} \sin\left(\theta + \frac{2\pi}{3}\right) \tag{1.53}$$

$$M_{fa} = \int_{\theta}^{\theta+\pi} \frac{N_f}{2} \sin(\beta - \theta) \left[ \int_{\beta-\pi}^{\beta} \mu_0 \frac{N_a}{2} \cos\phi \{a_0 + a_2 \cos(2\phi - 2\theta)\} rl\, d\phi \right] d\beta$$

$$= M_{af1} \cos\theta \tag{1.54}$$

$$M_{fb} = \int_{\theta}^{\theta+\pi} \frac{N_f}{2} \sin(\beta - \theta)$$

$$\times \left[ \int_{\beta-\pi}^{\beta} \mu_0 \frac{N_a}{2} \cos\left(\phi - \frac{2\pi}{3}\right) \{a_0 + a_2 \cos(2\phi - 2\theta)\} rl\, d\phi \right] d\beta$$

$$= M_{af1} \cos\left(\theta - \frac{2\pi}{3}\right) \tag{1.55}$$

$$M_{fc} = \int_{\theta}^{\theta+\pi} \frac{N_f}{2} \sin(\beta - \theta)$$

$$\times \left[ \int_{\beta-\pi}^{\beta} \mu_0 \frac{N_a}{2} \cos\left(\phi + \frac{2\pi}{3}\right) \{a_0 + a_2 \cos(2\phi - 2\theta)\} rl\, d\phi \right] d\beta$$

$$= M_{af1} \cos\left(\theta + \frac{2\pi}{3}\right) \tag{1.56}$$

$$L_f = L_{fl} + \int\limits_{\theta}^{\theta+\pi} \frac{N_f}{2} \sin(\beta - \theta)$$

$$\times \left[ \int\limits_{\beta-\pi}^{\beta} \mu_0 \frac{N_f}{2} \cos(\phi - \theta) \{a_0 + a_2 \cos(2\phi - 2\theta)\} rl \, d\phi \right] d\beta$$

$$= L_{fl} + \frac{N_f^2}{8} \pi \mu_0 rl \, (2a_0 + a_2) \tag{1.57}$$

$$M_{f1d} = \int\limits_{\theta}^{\theta+\pi} \frac{N_f}{2} \sin(\beta - \theta)$$

$$\times \left[ \int\limits_{\beta-\pi}^{\beta} \mu_0 \frac{N_{1d}}{2} \cos(\phi - \theta) \{a_0 + a_2 \cos(2\phi - 2\theta)\} rl \, d\phi \right] d\beta$$

$$= \frac{N_f N_{1d}}{8} \pi \mu_0 rl \, (2a_0 + a_2) \tag{1.58}$$

$$M_{f1q} = - \int\limits_{\theta}^{\theta+\pi} \frac{N_f}{2} \sin(\beta - \theta)$$

$$\times \left[ \int\limits_{\beta-\pi}^{\beta} \mu_0 \frac{N_{1q}}{2} \sin(\phi - \theta) \{a_0 + a_2 \cos(2\phi - 2\theta)\} rl \, d\phi \right] d\beta$$

$$= 0 \tag{1.59}$$

$$M_{f2q} = - \int\limits_{\theta}^{\theta+\pi} \frac{N_f}{2} \sin(\beta - \theta)$$

$$\times \left[ \int\limits_{\beta-\pi}^{\beta} \mu_0 \frac{N_{2q}}{2} \sin(\phi - \theta) \{a_0 + a_2 \cos(2\phi - 2\theta)\} rl \, d\phi \right] d\beta$$

$$= 0 \tag{1.60}$$

$$M_{1da} = \int\limits_{\theta}^{\theta+\pi} \frac{N_{1d}}{2} \sin(\beta - \theta) \left[ \int\limits_{\beta-\pi}^{\beta} \mu_0 \frac{N_a}{2} \cos\phi \{a_0 + a_2 \cos(2\phi - 2\theta)\} rl\, d\phi \right] d\beta$$

$$= M_{a1d1} \cos\theta \tag{1.61}$$

$$M_{1db} = \int\limits_{\theta}^{\theta+\pi} \frac{N_{1d}}{2} \sin(\beta - \theta)$$

$$\times \left[ \int\limits_{\beta-\pi}^{\beta} \mu_0 \frac{N_a}{2} \cos\left(\phi - \frac{2\pi}{3}\right) \{a_0 + a_2 \cos(2\phi - 2\theta)\} rl\, d\phi \right] d\beta$$

$$= M_{a1d1} \cos\left(\theta - \frac{2\pi}{3}\right) \tag{1.62}$$

$$M_{1dc} = \int\limits_{\theta}^{\theta+\pi} \frac{N_{1d}}{2} \sin(\beta - \theta)$$

$$\times \left[ \int\limits_{\beta-\pi}^{\beta} \mu_0 \frac{N_a}{2} \cos\left(\phi + \frac{2\pi}{3}\right) \{a_0 + a_2 \cos(2\phi - 2\theta)\} rl\, d\phi \right] d\beta$$

$$= M_{a1d1} \cos\left(\theta + \frac{2\pi}{3}\right) \tag{1.63}$$

$$M_{1df} = \int\limits_{\theta}^{\theta+\pi} \frac{N_{1d}}{2} \sin(\beta - \theta)$$

$$\times \left[ \int\limits_{\beta-\pi}^{\beta} \mu_0 \frac{N_f}{2} \cos(\phi - \theta) \{a_0 + a_2 \cos(2\phi - 2\theta)\} rl\, d\phi \right] d\beta$$

$$= \frac{N_f N_{1d}}{8} \pi \mu_0 rl\, (2a_0 + a_2) \tag{1.64}$$

$$L_{1d} = L_{1dl} + \int\limits_{\theta}^{\theta+\pi} \frac{N_{1d}}{2} \sin(\beta - \theta)$$

$$\times \left[ \int\limits_{\beta-\pi}^{\beta} \mu_0 \frac{N_{1d}}{2} \cos(\phi - \theta) \{a_0 + a_2 \cos(2\phi - 2\theta)\} \, rl \, d\phi \right] d\beta$$

$$= L_{1dl} + \frac{N_{1d}^2}{8} \pi \mu_0 rl \, (2a_0 + a_2) \tag{1.65}$$

$$M_{1d1q} = - \int\limits_{\theta}^{\theta+\pi} \frac{N_{1d}}{2} \sin(\beta - \theta)$$

$$\times \left[ \int\limits_{\beta-\pi}^{\beta} \mu_0 \frac{N_{1q}}{2} \sin(\phi - \theta) \{a_0 + a_2 \cos(2\phi - 2\theta)\} \, rl \, d\phi \right] d\beta$$

$$= 0 \tag{1.66}$$

$$M_{1d2q} = - \int\limits_{\theta}^{\theta+\pi} \frac{N_{1d}}{2} \sin(\beta - \theta)$$

$$\times \left[ \int\limits_{\beta-\pi}^{\beta} \mu_0 \frac{N_{2q}}{2} \sin(\phi - \theta) \{a_0 + a_2 \cos(2\phi - 2\theta)\} \, rl \, d\phi \right] d\beta$$

$$= 0 \tag{1.67}$$

$$M_{1qa} = \int\limits_{\theta-\pi/2}^{\theta+\pi/2} \frac{N_{1q}}{2} \cos(\beta - \theta)$$

$$\times \left[ \int\limits_{\beta-\pi}^{\beta} \mu_0 \frac{N_a}{2} \cos\phi \{a_0 + a_2 \cos(2\phi - 2\theta)\} \, rl \, d\phi \right] d\beta$$

$$= M_{a1q1} \sin\theta \tag{1.68}$$

$$M_{1qb} = \int_{\theta-\pi/2}^{\theta+\pi/2} \frac{N_{1q}}{2} \cos(\beta - \theta)$$

$$\times \left[ \int_{\beta-\pi}^{\beta} \mu_0 \frac{N_a}{2} \cos\left(\phi - \frac{2\pi}{3}\right) \{a_0 + a_2 \cos(2\phi - 2\theta)\} \, rl \, d\phi \right] d\beta$$

$$= M_{a1q1} \sin\left(\theta - \frac{2\pi}{3}\right) \tag{1.69}$$

$$M_{1qc} = \int_{\theta-\pi/2}^{\theta+\pi/2} \frac{N_{1q}}{2} \cos(\beta - \theta)$$

$$\times \left[ \int_{\beta-\pi}^{\beta} \mu_0 \frac{N_a}{2} \cos\left(\phi + \frac{2\pi}{3}\right) \{a_0 + a_2 \cos(2\phi - 2\theta)\} \, rl \, d\phi \right] d\beta$$

$$= M_{a1q1} \sin\left(\theta + \frac{2\pi}{3}\right) \tag{1.70}$$

$$M_{1qf} = \int_{\theta-\pi/2}^{\theta+\pi/2} \frac{N_{1q}}{2} \cos(\beta - \theta)$$

$$\times \left[ \int_{\beta-\pi}^{\beta} \mu_0 \frac{N_f}{2} \cos(\phi - \theta) \{a_0 + a_2 \cos(2\phi - 2\theta)\} \, rl \, d\phi \right] d\beta$$

$$= 0 \tag{1.71}$$

$$M_{1q1d} = \int_{\theta-\pi/2}^{\theta+\pi/2} \frac{N_{1q}}{2} \cos(\beta - \theta)$$

$$\times \left[ \int_{\beta-\pi}^{\beta} \mu_0 \frac{N_{1d}}{2} \cos(\phi - \theta) \{a_0 + a_2 \cos(2\phi - 2\theta)\} \, rl \, d\phi \right] d\beta$$

$$= 0 \tag{1.72}$$

$$L_{1q} = L_{1ql} - \int_{\theta-\pi/2}^{\theta+\pi/2} \frac{N_{1q}}{2} \cos(\beta - \theta)$$

$$\times \left[ \int_{\beta-\pi}^{\beta} \mu_0 \frac{N_{1q}}{2} \sin(\phi - \theta) \{a_0 + a_2 \cos(2\phi - 2\theta)\} rl \, d\phi \right] d\beta$$

$$= L_{1ql} + \frac{N_{1q}^2}{8} \pi \mu_0 rl (2a_0 - a_2) \tag{1.73}$$

$$M_{1q2q} = - \int_{\theta-\pi/2}^{\theta+\pi/2} \frac{N_{1q}}{2} \cos(\beta - \theta)$$

$$\times \left[ \int_{\beta-\pi}^{\beta} \mu_0 \frac{N_{2q}}{2} \sin(\phi - \theta) \{a_0 + a_2 \cos(2\phi - 2\theta)\} rl \, d\phi \right] d\beta$$

$$= \frac{N_{1q} N_{2q}}{8} \pi \mu_0 rl (2a_0 - a_2) \tag{1.74}$$

$$M_{2qa} = \int_{\theta-\pi/2}^{\theta+\pi/2} \frac{N_{2q}}{2} \cos(\beta - \theta)$$

$$\times \left[ \int_{\beta-\pi}^{\beta} \mu_0 \frac{N_a}{2} \cos\phi \{a_0 + a_2 \cos(2\phi - 2\theta)\} rl \, d\phi \right] d\beta$$

$$= M_{a2q1} \sin\theta \tag{1.75}$$

$$M_{2qb} = \int_{\theta-\pi/2}^{\theta+\pi/2} \frac{N_{2q}}{2} \cos(\beta - \theta)$$

$$\times \left[ \int_{\beta-\pi}^{\beta} \mu_0 \frac{N_a}{2} \cos\left(\phi - \frac{2\pi}{3}\right) \{a_0 + a_2 \cos(2\phi - 2\theta)\} rl \, d\phi \right] d\beta$$

$$= M_{a2q1} \sin\left(\theta - \frac{2\pi}{3}\right) \tag{1.76}$$

$$M_{2qc} = \int\limits_{\theta-\pi/2}^{\theta+\pi/2} \frac{N_{2q}}{2} \cos(\beta - \theta)$$

$$\times \left[ \int\limits_{\beta-\pi}^{\beta} \mu_0 \frac{N_a}{2} \cos\left(\phi + \frac{2\pi}{3}\right) \{a_0 + a_2 \cos(2\phi - 2\theta)\} \, rl \, d\phi \right] d\beta$$

$$= M_{a2q1} \sin\left(\theta + \frac{2\pi}{3}\right) \tag{1.77}$$

$$M_{2qf} = \int\limits_{\theta-\pi/2}^{\theta+\pi/2} \frac{N_{2q}}{2} \cos(\beta - \theta)$$

$$\times \left[ \int\limits_{\beta-\pi}^{\beta} \mu_0 \frac{N_f}{2} \cos(\phi - \theta) \{a_0 + a_2 \cos(2\phi - 2\theta)\} \, rl \, d\phi \right] d\beta$$

$$= 0 \tag{1.78}$$

$$M_{2q1d} = \int\limits_{\theta-\pi/2}^{\theta+\pi/2} \frac{N_{2q}}{2} \cos(\beta - \theta)$$

$$\times \left[ \int\limits_{\beta-\pi}^{\beta} \mu_0 \frac{N_{1d}}{2} \cos(\phi - \theta) \{a_0 + a_2 \cos(2\phi - 2\theta)\} \, rl \, d\phi \right] d\beta$$

$$= 0 \tag{1.79}$$

$$M_{2q1q} = - \int\limits_{\theta-\pi/2}^{\theta+\pi/2} \frac{N_{2q}}{2} \cos(\beta - \theta)$$

$$\times \left[ \int\limits_{\beta-\pi}^{\beta} \mu_0 \frac{N_{1q}}{2} \sin(\phi - \theta) \{a_0 + a_2 \cos(2\phi - 2\theta)\} \, rl \, d\phi \right] d\beta$$

$$= \frac{N_{1q} N_{2q}}{8} \pi \mu_0 rl \, (2a_0 - a_2) \tag{1.80}$$

$$L_{2q} = L_{2ql} - \int_{\theta-\pi/2}^{\theta+\pi/2} \frac{N_{2q}}{2} \cos(\beta - \theta)$$

$$\times \left[ \int_{\beta-\pi}^{\beta} \mu_0 \frac{N_{2q}}{2} \sin(\phi - \theta)\{a_0 + a_2 \cos(2\phi - 2\theta)\} rl \, d\phi \right] d\beta$$

$$= L_{2ql} + \frac{N_{2q}^2}{8} \pi \mu_0 rl (2a_0 - a_2) \tag{1.81}$$

where

$$M_{ab0} \triangleq -\frac{N_a^2}{8} \pi \mu_0 rl a_0 \tag{1.82}$$

$$M_{af1} \triangleq \frac{N_a N_f}{8} \pi \mu_0 rl (2a_0 + a_2) \tag{1.83}$$

$$M_{a1d1} \triangleq \frac{N_a N_{1d}}{8} \pi \mu_0 rl (2a_0 + a_2) \tag{1.84}$$

$$M_{a1q1} \triangleq \frac{N_a N_{1q}}{8} \pi \mu_0 rl (2a_0 - a_2) \tag{1.85}$$

$$M_{a2q1} \triangleq \frac{N_a N_{2q}}{8} \pi \mu_0 rl (2a_0 - a_2) \tag{1.86}$$

It is to be noted that the order of the subscripts in the notation for mutual inductances does not affect the expression for mutual inductance. For example, $M_{ab} = M_{ba}$.

In general, the number of field poles is $p_f$. If the total number of turns in a winding is $N$, the number of turns per field pole pair is $2N/p_f$. The inductance of the part of a winding with $2N/p_f$ turns is equal to $2/p_f$ times the expression derived above, if $\theta$ is the electrical angle. The total inductance is obtained by multiplying this expression by $p_f/2$. Hence, the inductance expressions derived above are valid even if $p_f > 2$. The mechanical angle $\theta_m$ is related to the electrical angle $\theta$ by

$$\theta = \frac{p_f}{2} \theta_m \tag{1.87}$$

### 1.2.3 Equations of Motion

If friction and windage losses are neglected, by Newton's law,

$$J \frac{d^2\theta_m}{dt^2} = T_m - T_e \tag{1.88}$$

where $J$ is the combined moment of inertia of the rotor and the prime mover, $T_m$ is the mechanical torque, $T_e$ is the electromagnetic or electrical torque, and $\theta_m$ is the position of the rotor in mechanical radians. Equation (1.88) in electrical angle is

$$\frac{2}{p_f} J \frac{d^2\theta}{dt^2} = T_m - T_e \tag{1.89}$$

Multiplying by $2/p_f$ gives

$$\left(\frac{2}{p_f}\right)^2 J \frac{d^2\theta}{dt^2} = \frac{2}{p_f} T_m - \frac{2}{p_f} T_e \tag{1.90}$$

Defining $J' \triangleq (2/p_f)^2 J$, $T'_m \triangleq 2T_m/p_f$, and $T'_e \triangleq 2T_e/p_f$ and substituting in (1.90) give

$$J' \frac{d^2\theta}{dt^2} = T'_m - T'_e \tag{1.91}$$

$J'$, $T'_m$, and $T'_e$ are the moment of inertia, mechanical torque, and electrical torque, respectively, of an equivalent synchronous generator with two field poles [1]. Equation (1.91) can be written as the following two first-order equations.

$$\frac{d\theta}{dt} = \omega \tag{1.92}$$

$$\frac{d\omega}{dt} = \frac{1}{J'}(T'_m - T'_e) \tag{1.93}$$

$\omega$ is the speed of the rotor in electrical radian per second.

### 1.2.4 Expression for Electrical Torque

Let $W_e$ be the total electrical energy supplied to the magnetic field of all windings. From the circuit diagram in Fig. 1.3,

$$\frac{dW_e}{dt} = i_a \frac{d\psi_a}{dt} + i_b \frac{d\psi_b}{dt} + i_c \frac{d\psi_c}{dt} + i_f \frac{d\psi_f}{dt} + i_{1d} \frac{d\psi_{1d}}{dt} + i_{1q} \frac{d\psi_{1q}}{dt} + i_{2q} \frac{d\psi_{2q}}{dt} \tag{1.94}$$

Let $W_f$ be the energy stored in the magnetic field of all windings.

$$W_f = \frac{1}{2}
\begin{bmatrix} i_a \\ i_b \\ i_c \\ i_f \\ i_{1d} \\ i_{1q} \\ i_{2q} \end{bmatrix}^T
\begin{bmatrix}
L_a & M_{ab} & M_{ac} & M_{af} & M_{a1d} & M_{a1q} & M_{a2q} \\
M_{ba} & L_b & M_{bc} & M_{bf} & M_{b1d} & M_{b1q} & M_{b2q} \\
M_{ca} & M_{cb} & L_c & M_{cf} & M_{c1d} & M_{c1q} & M_{c2q} \\
M_{fa} & M_{fb} & M_{fc} & L_f & M_{f1d} & M_{f1q} & M_{f2q} \\
M_{1da} & M_{1db} & M_{1dc} & M_{1df} & L_{1d} & M_{1d1q} & M_{1d2q} \\
M_{1qa} & M_{1qb} & M_{1qc} & M_{1qf} & M_{1q1d} & L_{1q} & M_{1q2q} \\
M_{2qa} & M_{2qb} & M_{2qc} & M_{2qf} & M_{2q1d} & M_{2q1q} & L_{2q}
\end{bmatrix}
\begin{bmatrix} i_a \\ i_b \\ i_c \\ i_f \\ i_{1d} \\ i_{1q} \\ i_{2q} \end{bmatrix}
\tag{1.95}$$

Let $W_m$ be the mechanical work done by the magnetic field.

$$\frac{dW_m}{dt} = -T_e\frac{d\theta_m}{dt} = -T_e'\frac{d\theta}{dt} \tag{1.96}$$

By the law of conservation of energy,

$$\frac{dW_f}{dt} = \frac{dW_e}{dt} - \frac{dW_m}{dt} \tag{1.97}$$

From (1.94) to (1.97),

$$T_e' = -\frac{1}{2}\begin{bmatrix} i_a \\ i_b \\ i_c \\ i_f \\ i_{1d} \\ i_{1q} \\ i_{2q} \end{bmatrix}^T \left(\frac{d}{d\theta}\begin{bmatrix} L_a & M_{ab} & M_{ac} & M_{af} & M_{a1d} & M_{a1q} & M_{a2q} \\ M_{ba} & L_b & M_{bc} & M_{bf} & M_{b1d} & M_{b1q} & M_{b2q} \\ M_{ca} & M_{cb} & L_c & M_{cf} & M_{c1d} & M_{c1q} & M_{c2q} \\ M_{fa} & M_{fb} & M_{fc} & L_f & M_{f1d} & M_{f1q} & M_{f2q} \\ M_{1da} & M_{1db} & M_{1dc} & M_{1df} & L_{1d} & M_{1d1q} & M_{1d2q} \\ M_{1qa} & M_{1qb} & M_{1qc} & M_{1qf} & M_{1q1d} & L_{1q} & M_{1q2q} \\ M_{2qa} & M_{2qb} & M_{2qc} & M_{2qf} & M_{2q1d} & M_{2q1q} & L_{2q} \end{bmatrix}\right)\begin{bmatrix} i_a \\ i_b \\ i_c \\ i_f \\ i_{1d} \\ i_{1q} \\ i_{2q} \end{bmatrix} \tag{1.98}$$

## 1.3  Park's Transformation

In the model obtained in the last section, there are inductances which are time-varying since they depend on $\theta$, and $\theta$ varies with time. The model can be simplified by transformation of stator variables. Let

$$\psi_s \triangleq \begin{bmatrix} \psi_a \\ \psi_b \\ \psi_c \end{bmatrix}, \quad i_s \triangleq \begin{bmatrix} i_a \\ i_b \\ i_c \end{bmatrix}, \quad v_s \triangleq \begin{bmatrix} v_a \\ v_b \\ v_c \end{bmatrix}, \quad \psi_r \triangleq \begin{bmatrix} \psi_f \\ \psi_{1d} \\ \psi_{1q} \\ \psi_{2q} \end{bmatrix}, \quad i_r \triangleq \begin{bmatrix} i_f \\ i_{1d} \\ i_{1q} \\ i_{2q} \end{bmatrix} \tag{1.99}$$

Equations (1.1)–(1.3) and (1.8) can be written as

$$\frac{d\psi_s}{dt} = -R_a i_s - v_s \tag{1.100}$$

$$\psi_s = L_s i_s + M_{sr} i_r \tag{1.101}$$

$$\psi_r = M_{sr}^T i_s + L_r i_r \tag{1.102}$$

where

$$L_s \triangleq \begin{bmatrix} L_{a0} & M_{ab0} & M_{ab0} \\ M_{ab0} & L_{a0} & M_{ab0} \\ M_{ab0} & M_{ab0} & L_{a0} \end{bmatrix} + L_{a2}\begin{bmatrix} \cos(2\theta) & \cos(2\theta - 2\pi/3) & \cos(2\theta + 2\pi/3) \\ \cos(2\theta - 2\pi/3) & \cos(2\theta + 2\pi/3) & \cos(2\theta) \\ \cos(2\theta + 2\pi/3) & \cos(2\theta) & \cos(2\theta - 2\pi/3) \end{bmatrix} \tag{1.103}$$

$$M_{sr} \triangleq \begin{bmatrix} M_{af1}\cos\theta & M_{a1d1}\cos\theta \\ M_{af1}\cos(\theta - 2\pi/3) & M_{a1d1}\cos(\theta - 2\pi/3) \\ M_{af1}\cos(\theta + 2\pi/3) & M_{a1d1}\cos(\theta + 2\pi/3) \end{bmatrix}$$

$$\begin{bmatrix} M_{a1q1}\sin\theta & M_{a2q1}\sin\theta \\ M_{a1q1}\sin(\theta - 2\pi/3) & M_{a2q1}\sin(\theta - 2\pi/3) \\ M_{a1q1}\sin(\theta + 2\pi/3) & M_{a2q1}\sin(\theta + 2\pi/3) \end{bmatrix} \qquad (1.104)$$

$$L_r \triangleq \begin{bmatrix} L_f & M_{f1d} & 0 & 0 \\ M_{f1d} & L_{1d} & 0 & 0 \\ 0 & 0 & L_{1q} & M_{1q2q} \\ 0 & 0 & M_{1q2q} & L_{2q} \end{bmatrix} \qquad (1.105)$$

Let $\psi_s$, $i_s$, and $v_s$ be transformed to $\psi_{dq0}$, $i_{dq0}$, and $v_{dq0}$, respectively, as follows:

$$\psi_{dq0} \triangleq T_P \psi_s, \quad i_{dq0} \triangleq T_P i_s, \quad v_{dq0} \triangleq T_P v_s \qquad (1.106)$$

where $T_P$ is a $3 \times 3$ nonsingular matrix and

$$\psi_{dq0} \triangleq \begin{bmatrix} \psi_d \\ \psi_q \\ \psi_0 \end{bmatrix}, \quad i_{dq0} \triangleq \begin{bmatrix} i_d \\ i_q \\ i_0 \end{bmatrix}, \quad v_{dq0} \triangleq \begin{bmatrix} v_d \\ v_q \\ v_0 \end{bmatrix} \qquad (1.107)$$

Using this transformation in (1.101) gives

$$T_P^{-1}\psi_{dq0} = L_s T_P^{-1} i_{dq0} + M_{sr} i_r \qquad (1.108)$$

Pre-multiplying by $T_P$ gives

$$\psi_{dq0} = T_P L_s T_P^{-1} i_{dq0} + T_P M_{sr} i_r \qquad (1.109)$$

Let $T_P$ be chosen such that $T_P L_s T_P^{-1}$ is a diagonal matrix [5]; then, the transformation is known as Park's transformation. The columns of $T_P^{-1}$ are right eigenvectors of $L_s$. $T_P$ is given by

$$T_P^{-1} = \begin{bmatrix} k_d \cos\theta & k_q \sin\theta & k_0 \\ k_d \cos(\theta - 2\pi/3) & k_q \sin(\theta - 2\pi/3) & k_0 \\ k_d \cos(\theta + 2\pi/3) & k_q \sin(\theta + 2\pi/3) & k_0 \end{bmatrix} \qquad (1.110)$$

where $k_d$, $k_q$, and $k_0$ can be chosen arbitrarily. The diagonal elements of $T_P L_s T_P^{-1}$ are the eigenvalues of $L_s$.

The power at the synchronous generator terminals is

$$P = v_s^T i_s = v_{dq0}^T \left(T_P^{-1}\right)^T T_P^{-1} i_{dq0} \qquad (1.111)$$

The matrix $\left(T_P^{-1}\right)^T T_P^{-1}$ is diagonal, given by

$$\left(T_P^{-1}\right)^T T_P^{-1} = \begin{bmatrix} 3k_d^2/2 & 0 & 0 \\ 0 & 3k_q^2/2 & 0 \\ 0 & 0 & 3k_0^2 \end{bmatrix} \tag{1.112}$$

The transformation is said to be power invariant if $P = v_{dq0}^T i_{dq0}$ [1, 2]. Power invariance is satisfied if $T_P^{-1} = T_P^T$. The values of $k_d$, $k_q$, and $k_0$ for power invariance are

$$k_d = \pm\sqrt{\frac{2}{3}}, k_q = \pm\sqrt{\frac{2}{3}}, k_0 = \pm\sqrt{\frac{1}{3}} \tag{1.113}$$

If positive values are used,

$$T_P = \frac{1}{\sqrt{3}} \begin{bmatrix} \sqrt{2}\cos\theta & \sqrt{2}\cos(\theta - 2\pi/3) & \sqrt{2}\cos(\theta + 2\pi/3) \\ \sqrt{2}\sin\theta & \sqrt{2}\sin(\theta - 2\pi/3) & \sqrt{2}\sin(\theta + 2\pi/3) \\ 1 & 1 & 1 \end{bmatrix} \tag{1.114}$$

Using this $T_P$ in (1.109) gives

$$\begin{bmatrix} \psi_d \\ \psi_q \\ \psi_0 \end{bmatrix} = \begin{bmatrix} L_d & 0 & 0 \\ 0 & L_q & 0 \\ 0 & 0 & L_0 \end{bmatrix} \begin{bmatrix} i_d \\ i_q \\ i_0 \end{bmatrix} + \begin{bmatrix} M_{df} & M_{d1d} & 0 & 0 \\ 0 & 0 & M_{q1q} & M_{q2q} \\ 0 & 0 & 0 & 0 \end{bmatrix} \begin{bmatrix} i_f \\ i_{1d} \\ i_{1q} \\ i_{2q} \end{bmatrix} \tag{1.115}$$

where

$$L_d \triangleq L_{a0} - M_{ab0} + \frac{3}{2}L_{a2}, \quad L_q \triangleq L_{a0} - M_{ab0} - \frac{3}{2}L_{a2}, \quad L_0 \triangleq L_{a0} + 2M_{ab0} \tag{1.116}$$

$$M_{df} \triangleq \sqrt{\frac{3}{2}}M_{af1}, \quad M_{d1d} \triangleq \sqrt{\frac{3}{2}}M_{a1d1}, \quad M_{q1q} \triangleq \sqrt{\frac{3}{2}}M_{a1q1}, \quad M_{q2q} \triangleq \sqrt{\frac{3}{2}}M_{a2q1} \tag{1.117}$$

$L_d$, $L_q$, and $L_0$ are called $d$-axis inductance, $q$-axis inductance, and zero sequence inductance, respectively. From (1.102) and (1.106),

$$\begin{bmatrix} \psi_f \\ \psi_{1d} \\ \psi_{1q} \\ \psi_{2q} \end{bmatrix} = \begin{bmatrix} M_{df} & 0 & 0 \\ M_{d1d} & 0 & 0 \\ 0 & M_{q1q} & 0 \\ 0 & M_{q2q} & 0 \end{bmatrix} \begin{bmatrix} i_d \\ i_q \\ i_0 \end{bmatrix} + \begin{bmatrix} L_f & M_{f1d} & 0 & 0 \\ M_{f1d} & L_{1d} & 0 & 0 \\ 0 & 0 & L_{1q} & M_{1q2q} \\ 0 & 0 & M_{1q2q} & L_{2q} \end{bmatrix} \begin{bmatrix} i_f \\ i_{1d} \\ i_{1q} \\ i_{2q} \end{bmatrix} \tag{1.118}$$

From (1.100) and (1.106),

$$\frac{\mathrm{d}\left(T_P^T \psi_{dq0}\right)}{\mathrm{d}t} = -R_a T_P^T i_{dq0} - T_P^T v_{dq0} \tag{1.119}$$

Pre-multiplying by $T_P$ gives

$$\frac{\mathrm{d}\psi_{dq0}}{\mathrm{d}t} = -\omega M \psi_{dq0} - R_a i_{dq0} - v_{dq0} \tag{1.120}$$

where

$$M \triangleq \begin{bmatrix} 0 & 1 & 0 \\ -1 & 0 & 0 \\ 0 & 0 & 0 \end{bmatrix} \tag{1.121}$$

From (1.98),

$$T_e' = -\frac{1}{2}\begin{bmatrix} i_s^T & i_r^T \end{bmatrix}\begin{bmatrix} \mathrm{d}L_s/\mathrm{d}\theta & \mathrm{d}M_{sr}/\mathrm{d}\theta \\ \mathrm{d}M_{sr}^T/\mathrm{d}\theta & O \end{bmatrix}\begin{bmatrix} i_s \\ i_r \end{bmatrix} \tag{1.122}$$

where $O$ is a null matrix. Substituting $T_P^T i_{dq0}$ for $i_s$ in (1.122) and simplifying give

$$T_e' = -\frac{1}{2}i_{dq0}^T T_P \frac{\mathrm{d}L_s}{\mathrm{d}\theta} T_P^T i_{dq0} - i_{dq0}^T T_P \frac{\mathrm{d}M_{sr}}{\mathrm{d}\theta} i_r = \psi_d i_q - \psi_q i_d \tag{1.123}$$

If $\omega_o$ is the steady-state or operating speed, then

$$\theta = \omega_o t + \delta \tag{1.124}$$

$\delta$ is the angular position of the rotor in electrical radian with respect to a reference rotating at speed $\omega_o$.

With the transformation of stator variables and use of $\delta$ instead of $\theta$, the equations governing the synchronous generator are

$$\frac{\mathrm{d}\psi_d}{\mathrm{d}t} = -\omega\psi_q - R_a i_d - v_d \tag{1.125}$$

$$\frac{\mathrm{d}\psi_q}{\mathrm{d}t} = \omega\psi_d - R_a i_q - v_q \tag{1.126}$$

$$\frac{\mathrm{d}\psi_0}{\mathrm{d}t} = -R_a i_0 - v_0 \tag{1.127}$$

$$\frac{\mathrm{d}\psi_f}{\mathrm{d}t} = -R_f i_f + v_f \tag{1.128}$$

$$\frac{\mathrm{d}\psi_{1d}}{\mathrm{d}t} = -R_{1d} i_{1d} \tag{1.129}$$

$$\frac{d\psi_{1q}}{dt} = -R_{1q}i_{1q} \tag{1.130}$$

$$\frac{d\psi_{2q}}{dt} = -R_{2q}i_{2q} \tag{1.131}$$

$$\frac{d\delta}{dt} = \omega - \omega_o \tag{1.132}$$

$$\frac{d\omega}{dt} = \frac{1}{J'}\left(T'_m - \psi_d i_q + \psi_q i_d\right) \tag{1.133}$$

$$\psi_d = L_d i_d + M_{df}i_f + M_{d1d}i_{1d} \tag{1.134}$$

$$\psi_q = L_q i_q + M_{q1q}i_{1q} + M_{q2q}i_{2q} \tag{1.135}$$

$$\psi_f = M_{df}i_d + L_{f}i_f + M_{f1d}i_{1d} \tag{1.136}$$

$$\psi_{1d} = M_{d1d}i_d + M_{f1d}i_f + L_{1d}i_{1d} \tag{1.137}$$

$$\psi_{1q} = M_{q1q}i_q + L_{1q}i_{1q} + M_{1q2q}i_{2q} \tag{1.138}$$

$$\psi_{2q} = M_{q2q}i_q + M_{1q2q}i_{1q} + L_{2q}i_{2q} \tag{1.139}$$

$$\psi_0 = L_0 i_0 \tag{1.140}$$

Park's transformation results in constant inductances. It is apparent that Park's transformation results in the replacement of the stator windings by three windings whose voltage, current, and flux linkage are $v_{dq0}$, $i_{dq0}$, and $\psi_{dq0}$, respectively. From (1.134) to (1.139), it is apparent that one winding is on the $d$-axis and one winding is on the $q$-axis; hence, the voltage, current, and flux linkage have notations with the subscripts $d$ and $q$. However, the presence of the first term on the right-hand side of (1.125) and (1.126) suggests that this is not true [6]. The quantities having a notation with the subscript 0 are zero sequence quantities.

## 1.4 Transformation of Rotor Variables

In order to obtain a model whose parameters can be determined, the model obtained in the previous section is partitioned into three parts as shown in Fig. 1.8.

An equivalent synchronous generator model is shown in Fig. 1.9. Parts 2 and 3 in Fig. 1.8 are replaced by the following equations in Fig. 1.9.

$$I_d(s) = G_1(s)\Psi_d(s) + G_2(s)V_f(s) \tag{1.141}$$

$$I_q(s) = G_3(s)\Psi_q(s) \tag{1.142}$$

where $I_d(s)$, $\Psi_d(s)$, $V_f(s)$, $I_q(s)$, and $\Psi_q(s)$ are Laplace transform of $i_d$, $\psi_d$, $v_f$, $i_q$, and $\psi_q$, respectively, and $G_1(s)$, $G_2(s)$, and $G_3(s)$ are transfer functions. The following three transfer functions are derived below.

$$G_1(s) = \left.\frac{I_d(s)}{\Psi_d(s)}\right|_{v_f=0} \tag{1.143}$$

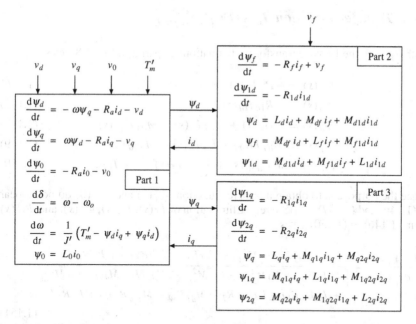

**Fig. 1.8** Partition of the synchronous generator model

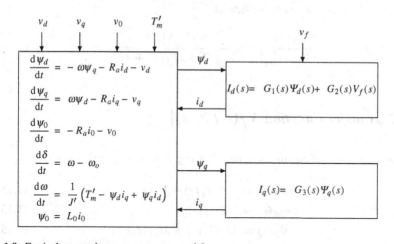

**Fig. 1.9** Equivalent synchronous generator model

$$-\frac{G_1(s)}{G_2(s)} = \left.\frac{V_f(s)}{\Psi_d(s)}\right|_{i_d=0} \tag{1.144}$$

$$G_3(s) = \frac{I_q(s)}{\Psi_q(s)} \tag{1.145}$$

## 1.4.1 Transfer Function $I_d(s)/\Psi_d(s)|_{v_f=0}$

With $v_f = 0$, the Laplace transform of equations in part 2 of Fig. 1.8 gives

$$s\Psi_f(s) = -R_f I_f(s) \tag{1.146}$$

$$s\Psi_{1d}(s) = -R_{1d} I_{1d}(s) \tag{1.147}$$

$$\Psi_d(s) = L_d I_d(s) + M_{df} I_f(s) + M_{d1d} I_{1d}(s) \tag{1.148}$$

$$\Psi_f(s) = M_{df} I_d(s) + L_f I_f(s) + M_{f1d} I_{1d}(s) \tag{1.149}$$

$$\Psi_{1d}(s) = M_{d1d} I_d(s) + M_{f1d} I_f(s) + L_{1d} I_{1d}(s) \tag{1.150}$$

Since the purpose is to obtain a transfer function, $\psi_f(0)$ and $\psi_{1d}(0)$ do not appear in (1.146) and (1.147), respectively. Elimination of $I_f(s)$, $I_{1d}(s)$, $\Psi_f(s)$, and $\Psi_{1d}(s)$ from (1.146) to (1.150) gives

$$\left.\frac{I_d(s)}{\Psi_d(s)}\right|_{v_f=0} = \frac{L_f L_{1d} - M_{f1d}^2 + (L_f R_{1d} + L_{1d} R_f)/s + R_f R_{1d}/s^2}{\begin{bmatrix} L_d L_f L_{1d} - L_d M_{f1d}^2 - L_{1d} M_{df}^2 + 2M_{df} M_{f1d} M_{d1d} - L_f M_{d1d}^2 \\ +(L_d L_f R_{1d} + L_d L_{1d} R_f - M_{df}^2 R_{1d} - M_{d1d}^2 R_f)/s + L_d R_f R_{1d}/s^2 \end{bmatrix}} \tag{1.151}$$

This transfer function can be written as

$$\left.\frac{I_d(s)}{\Psi_d(s)}\right|_{v_f=0} = \frac{\left(1 + s T_{do}'\right)\left(1 + s T_{do}''\right)}{L_d\left(1 + s T_d'\right)\left(1 + s T_d''\right)} \tag{1.152}$$

## 1.4.2 Transfer Function $V_f(s)/\Psi_d(s)|_{i_d=0}$

Laplace transform of equations in part 2 of Fig. 1.8, with $i_d = 0$, gives

$$s\Psi_f(s) = -R_f I_f(s) + V_f(s) \tag{1.153}$$

$$s\Psi_{1d}(s) = -R_{1d} I_{1d}(s) \tag{1.154}$$

$$\Psi_d(s) = M_{df} I_f(s) + M_{d1d} I_{1d}(s) \tag{1.155}$$

$$\Psi_f(s) = L_f I_f(s) + M_{f1d} I_{1d}(s) \tag{1.156}$$

$$\Psi_{1d}(s) = M_{f1d} I_f(s) + L_{1d} I_{1d}(s) \tag{1.157}$$

Elimination of $I_f(s)$, $I_{1d}(s)$, $\Psi_f(s)$, and $\Psi_{1d}(s)$ from (1.153) to (1.157) gives

$$\left.\frac{V_f(s)}{\Psi_d(s)}\right|_{i_d=0} = \frac{L_f L_{1d} - M_{f1d}^2 + (L_f R_{1d} + L_{1d} R_f)/s + R_f R_{1d}/s^2}{(M_{df} L_{1d} - M_{d1d} M_{f1d})/s + M_{df} R_{1d}/s^2} \tag{1.158}$$

This can be written as

$$\frac{V_f(s)}{\Psi_d(s)}\bigg|_{i_d=0} = \frac{R_f\left(1 + sT'_{do}\right)\left(1 + sT''_{do}\right)}{M_{df}\left(1 + sT''_{dc}\right)} \tag{1.159}$$

### 1.4.3 Transfer Function $I_q(s)/\Psi_q(s)$

Laplace transform of equations in part 3 of Fig. 1.8 gives

$$s\Psi_{1q}(s) = -R_{1q}I_{1q}(s) \tag{1.160}$$

$$s\Psi_{2q}(s) = -R_{2q}I_{2q}(s) \tag{1.161}$$

$$\Psi_q(s) = L_qI_q(s) + M_{q1q}I_{1q}(s) + M_{q2q}I_{2q}(s) \tag{1.162}$$

$$\Psi_{1q}(s) = M_{q1q}I_q(s) + L_{1q}I_{1q}(s) + M_{1q2q}I_{2q}(s) \tag{1.163}$$

$$\Psi_{2q}(s) = M_{q2q}I_q(s) + M_{1q2q}I_{1q}(s) + L_{2q}I_{2q}(s) \tag{1.164}$$

Elimination of $I_{1q}(s)$, $I_{2q}(s)$, $\Psi_{1q}(s)$, and $\Psi_{2q}(s)$ from (1.160) to (1.164) gives

$$\frac{I_q(s)}{\Psi_q(s)} = \frac{L_{1q}L_{2q} - M_{1q2q}^2 + (L_{1q}R_{2q} + L_{2q}R_{1q})/s + R_{1q}R_{2q}/s^2}{\left[\begin{array}{c} L_qL_{1q}L_{2q} - L_qM_{1q2q}^2 - L_{2q}M_{q1q}^2 + 2M_{q1q}M_{1q2q}M_{q2q} - L_{1q}M_{q2q}^2 \\ +(L_qL_{1q}R_{2q} + L_qL_{2q}R_{1q} - R_{2q}M_{q1q}^2 - R_{1q}M_{q2q}^2)/s + L_qR_{1q}R_{2q}/s^2 \end{array}\right]} \tag{1.165}$$

This can be written as

$$\frac{I_q(s)}{\Psi_q(s)} = \frac{\left(1 + sT'_{qo}\right)\left(1 + sT''_{qo}\right)}{L_q\left(1 + sT'_q\right)\left(1 + sT''_q\right)} \tag{1.166}$$

## 1.5 Tests for the Determination of Parameters

The parameters in (1.152) and (1.159) are called $d$-axis parameters, and those in (1.166) are called $q$-axis parameters. These parameters can be determined by tests. This section describes one type of tests known as standstill frequency response tests in which the rotor is standstill at a certain position.

### 1.5.1 Determination of d-Axis Parameters

There are two tests to be conducted to determine the $d$-axis parameters. The circuit diagram for the first test is shown in Fig. 1.10 [1, 6]. The rotor position should be

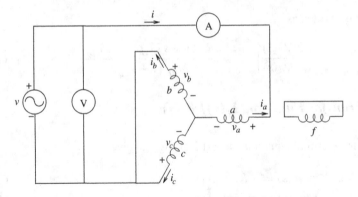

**Fig. 1.10** Circuit diagram for the first test for the determination of $d$-axis parameters

such that $\theta = 0$. Since the rotor is at standstill, $\omega = 0$. From the circuit diagram in Fig. 1.10,

$$v = v_a - v_b \tag{1.167}$$

$$v_b = v_c \tag{1.168}$$

$$i = -i_a = i_b + i_c \tag{1.169}$$

$$v_f = 0 \tag{1.170}$$

Hence,

$$v_d = \sqrt{\frac{2}{3}}v \tag{1.171}$$

$$i_d = -\sqrt{\frac{3}{2}}i \tag{1.172}$$

$$\frac{I_d(s)}{V_d(s)} = -\frac{3}{2}\frac{I(s)}{V(s)} \tag{1.173}$$

Since $\omega = 0$, from (1.125),

$$\frac{d\psi_d}{dt} = -R_a i_d - v_d \tag{1.174}$$

Laplace transform of this equation gives

$$\left.\frac{\Psi_d(s)}{I_d(s)}\right|_{v_f=0} = -\frac{V_d(s)}{sI_d(s)} - \frac{R_a}{s} \tag{1.175}$$

From (1.173) and (1.175),

**Fig. 1.11** Circuit diagram for the second test for the determination of $d$-axis parameters

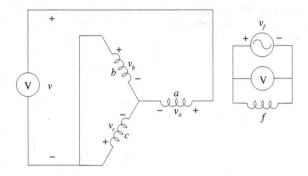

$$\left. \frac{I_d(s)}{\Psi_d(s)} \right|_{v_f=0} = \frac{1}{[2/(3s)]V(s)/I(s) - R_a/s} \tag{1.176}$$

The magnitude and phase angle of this transfer function are obtained at different frequencies.

The circuit diagram for the second test is shown in Fig. 1.11 [6]. The rotor position should be such that $\theta = 0$. From the circuit diagram in Fig. 1.11,

$$v_d = \sqrt{\frac{2}{3}}v \tag{1.177}$$

$$i_d = 0 \tag{1.178}$$

Since $\omega = 0$ and $i_d = 0$, from (1.125),

$$\frac{d\psi_d}{dt} = -v_d \tag{1.179}$$

From (1.177) and (1.179),

$$\left. \frac{V_f(s)}{\Psi_d(s)} \right|_{i_d=0} = -\sqrt{\frac{3}{2}}s\frac{V_f(s)}{V(s)} \tag{1.180}$$

The magnitude and phase angle of this transfer function are obtained at different frequencies. From the values of the magnitude and the phase angle of the two transfer functions $I_d(s)/\Psi_d(s)|_{v_f=0}$ and $V_f(s)/\Psi_d(s)|_{i_d=0}$ at different frequencies, the values of the $d$-axis parameters $L_d$, $T'_{do}$, $T''_{do}$, $T'_d$, $T''_d$, $R_f/M_{df}$, and $T''_{dc}$ are estimated. It can be verified from the tests that $T'_d$, $T''_d$, $T'_{do}$, and $T''_{do}$ are real and positive, and $T''_{dc}$ is positive. The notations are such that $T'_d > T''_d$ and $T'_{do} > T''_{do}$.

**Fig. 1.12** Circuit diagram for
the test for the determination
of $q$-axis parameters

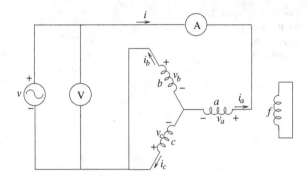

### 1.5.2 Determination of q-Axis Parameters

The $q$-axis parameters are determined by conducting a test, the circuit diagram for
which is shown in Fig. 1.12 [1, 6]. The rotor position should be such that $\theta = 90°$.
Since the rotor is at standstill, $\omega = 0$. From the circuit diagram in Fig. 1.12,

$$v_q = \sqrt{\frac{2}{3}}v \tag{1.181}$$

$$i_q = -\sqrt{\frac{3}{2}}i \tag{1.182}$$

$$\frac{V_q(s)}{I_q(s)} = -\frac{2}{3}\frac{V(s)}{I(s)} \tag{1.183}$$

Since $\omega = 0$, from (1.126),

$$\frac{d\psi_q}{dt} = -R_a i_q - v_q \tag{1.184}$$

Laplace transform of this equation gives

$$\frac{\Psi_q(s)}{I_q(s)} = -\frac{V_q(s)}{sI_q(s)} - \frac{R_a}{s} \tag{1.185}$$

From (1.183) and (1.185),

$$\frac{I_q(s)}{\Psi_q(s)} = \frac{1}{[2/(3s)]V(s)/I(s) - R_a/s} \tag{1.186}$$

From the values of the magnitude and the phase angle of this transfer function at
different frequencies, the values of the $q$-axis parameters $L_q, T'_{qo}, T''_{qo}, T'_q$, and $T''_q$

are estimated. It can be verified from the test that $T_q'$, $T_q''$, $T_{qo}'$, and $T_{qo}''$ are real and positive. The notations are such that $T_q' > T_q''$ and $T_{qo}' > T_{qo}''$.

## 1.6  Time Domain Model with Standard Parameters

From (1.141) to (1.145), (1.152), (1.159), and (1.166),

$$I_d(s) = \frac{(1 + sT_{do}')(1 + sT_{do}'')}{L_d(1 + sT_d')(1 + sT_d'')}\Psi_d(s) - \frac{M_{df}(1 + sT_{dc}'')}{R_f L_d(1 + sT_d')(1 + sT_d'')}V_f(s)$$

(1.187)

$$I_q(s) = \frac{(1 + sT_{qo}')(1 + sT_{qo}'')}{L_q(1 + sT_q')(1 + sT_q'')}\Psi_q(s)$$

(1.188)

By partial fraction expansion,

$$I_d(s) = \left(r_1 + \frac{r_2}{1 + sT_d'} + \frac{r_3}{1 + sT_d''}\right)\Psi_d(s) + \frac{M_{df}}{R_f}\left(\frac{r_4}{1 + sT_d'} + \frac{r_5}{1 + sT_d''}\right)V_f(s)$$

(1.189)

$$I_q(s) = \left(r_6 + \frac{r_7}{1 + sT_q'} + \frac{r_8}{1 + sT_q''}\right)\Psi_q(s)$$

(1.190)

where

$$r_1 \triangleq \frac{T_{do}'T_{do}''}{L_d T_d'T_d''}, \quad r_2 \triangleq \frac{(T_d' - T_{do}')(T_d' - T_{do}'')}{L_d T_d'(T_d' - T_d'')}, \quad r_3 \triangleq \frac{(T_d'' - T_{do}')(T_d'' - T_{do}'')}{L_d T_d''(T_d'' - T_d')}$$  (1.191)

$$r_4 \triangleq \frac{T_d' - T_{dc}''}{L_d(T_d'' - T_d')}, \quad r_5 \triangleq \frac{T_{dc}'' - T_d''}{L_d(T_d'' - T_d')}$$  (1.192)

$$r_6 \triangleq \frac{T_{qo}'T_{qo}''}{L_q T_q'T_q''}, \quad r_7 \triangleq \frac{(T_q' - T_{qo}')(T_q' - T_{qo}'')}{L_q T_q'(T_q' - T_q'')}, \quad r_8 \triangleq \frac{(T_q'' - T_{qo}')(T_q'' - T_{qo}'')}{L_q T_q''(T_q'' - T_q')}$$  (1.193)

Equations (1.189) and (1.190) can be put in the block diagram form shown in Figs. 1.13 and 1.14 where

$$E_f \triangleq \frac{M_{df}}{R_f}v_f$$

(1.194)

The time domain model is obtained with $\psi_F$, $\psi_{1D}$, $\psi_{1Q}$, and $\psi_{2Q}$ as the state variables associated with the rotor windings. The time domain model will be obtained in terms of the inductances $L_d$ and $L_q$ and other inductances defined below, instead of the residues $r_1$ to $r_8$.

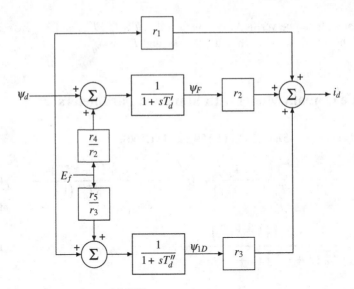

**Fig. 1.13**  Block diagram form of (1.189)

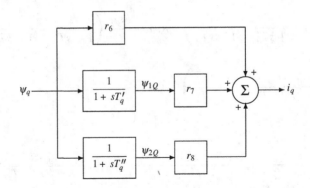

**Fig. 1.14**  Block diagram form of (1.190)

$L_d$ and $L_q$ are related to the residues by

$$L_d = \frac{1}{r_1 + r_2 + r_3}, \quad L_q = \frac{1}{r_6 + r_7 + r_8} \tag{1.195}$$

An inductance $L_d''$ called $d$-axis subtransient inductance and an inductance $L_q''$ called $q$-axis subtransient inductance are defined [4] as

$$L_d'' \triangleq \lim_{s \to \infty} \frac{\Psi_d(s)}{I_d(s)}\bigg|_{v_f = 0}, \quad L_q'' \triangleq \lim_{s \to \infty} \frac{\Psi_q(s)}{I_q(s)} \tag{1.196}$$

From (1.189), (1.190), and (1.196),

$$L''_d = \frac{1}{r_1}, \quad L''_q = \frac{1}{r_6} \tag{1.197}$$

An inductance $L'_d$ called $d$-axis transient inductance and an inductance $L'_q$ called $q$-axis transient inductance are defined as

$$L'_d \triangleq \frac{1}{r_1 + r_3}, \quad L'_q \triangleq \frac{1}{r_6 + r_8} \tag{1.198}$$

These inductances satisfy the inequality relations: $L_d > L'_d > L''_d > 0$ and $L_q > L'_q > L''_q > 0$.

$T'_d$, $T''_d$, $T'_q$, and $T''_q$ are called short-circuit time constants. $T'_{do}$, $T''_{do}$, $T'_{qo}$, and $T''_{qo}$ are called open-circuit time constants. These time constants and the inductances $L_d$, $L'_d$, $L''_d$, $L_q$, $L'_q$, and $L''_q$ are called standard parameters [2, 6]. The synchronous generator model with standard parameters is given by the following equations:

$$\frac{d\psi_d}{dt} = -\omega\psi_q - R_a i_d - v_d \tag{1.199}$$

$$\frac{d\psi_q}{dt} = \omega\psi_d - R_a i_q - v_q \tag{1.200}$$

$$\frac{d\psi_0}{dt} = -R_a i_0 - v_0 \tag{1.201}$$

$$\frac{d\psi_F}{dt} = \frac{1}{T'_d}\left[-\psi_F + \psi_d + \frac{L'_d(T''_{dc} - T'_d)}{(L'_d - L_d)(T'_d - T''_d)}E_f\right] \tag{1.202}$$

$$\frac{d\psi_{1D}}{dt} = \frac{1}{T''_d}\left[-\psi_{1D} + \psi_d + \frac{L'_d L''_d(T''_{dc} - T''_d)}{L_d(L''_d - L'_d)(T''_d - T'_d)}E_f\right] \tag{1.203}$$

$$\frac{d\psi_{1Q}}{dt} = \frac{1}{T'_q}\left(-\psi_{1Q} + \psi_q\right) \tag{1.204}$$

$$\frac{d\psi_{2Q}}{dt} = \frac{1}{T''_q}\left(-\psi_{2Q} + \psi_q\right) \tag{1.205}$$

$$\frac{d\delta}{dt} = \omega - \omega_0 \tag{1.206}$$

$$\frac{d\omega}{dt} = \frac{1}{J'}\left(T'_m - \psi_d i_q + \psi_q i_d\right) \tag{1.207}$$

$$i_d = \frac{1}{L''_d}\psi_d + \left(\frac{1}{L_d} - \frac{1}{L'_d}\right)\psi_F + \left(\frac{1}{L'_d} - \frac{1}{L''_d}\right)\psi_{1D} \tag{1.208}$$

$$i_q = \frac{1}{L''_q}\psi_q + \left(\frac{1}{L_q} - \frac{1}{L'_q}\right)\psi_{1Q} + \left(\frac{1}{L'_q} - \frac{1}{L''_q}\right)\psi_{2Q} \tag{1.209}$$

$$i_0 = \frac{1}{L_0}\psi_0 \tag{1.210}$$

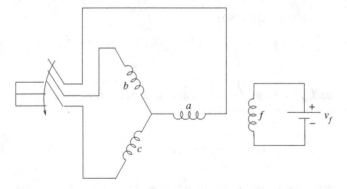

**Fig. 1.15**  Three-phase short circuit at the synchronous generator terminals

## 1.7  Time Constants

### 1.7.1  Short-Circuit Time Constants

$T_d'$, $T_d''$, $T_q'$, and $T_q''$ are called short-circuit time constants since these are the time constants during short circuit at the synchronous generator terminals. Suppose the synchronous generator is run at $\omega = \omega_o$, and then, its terminals are shorted by closing the switch shown in Fig. 1.15. During this condition, $v_d = v_q = v_0 = i_0 = 0$. If $R_a$ is neglected, the equations governing the synchronous generator during this condition are

$$\frac{d\psi_d}{dt} = -\omega_o \psi_q \tag{1.211}$$

$$\frac{d\psi_q}{dt} = \omega_o \psi_d \tag{1.212}$$

$$\frac{d\psi_F}{dt} = \frac{1}{T_d'}\left[-\psi_F + \psi_d + \frac{L_d'(T_{dc}'' - T_d')}{(L_d' - L_d)(T_d' - T_d'')}E_f\right] \tag{1.213}$$

$$\frac{d\psi_{1D}}{dt} = \frac{1}{T_d''}\left[-\psi_{1D} + \psi_d + \frac{L_d'L_d''(T_{dc}'' - T_d'')}{L_d(L_d'' - L_d')(T_d'' - T_d')}E_f\right] \tag{1.214}$$

$$\frac{d\psi_{1Q}}{dt} = \frac{1}{T_q'}\left(-\psi_{1Q} + \psi_q\right) \tag{1.215}$$

$$\frac{d\psi_{2Q}}{dt} = \frac{1}{T_q''}\left(-\psi_{2Q} + \psi_q\right) \tag{1.216}$$

The eigenvalues of the system during this condition are $-1/T_d'$, $-1/T_d''$, $-1/T_q'$, $-1/T_q''$, $j\omega_o$, $-j\omega_o$. Hence, $T_d'$, $T_d''$, $T_q'$, and $T_q''$ are called short-circuit time constants.

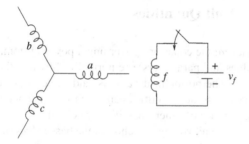

**Fig. 1.16** Field excited under open-circuit condition

## 1.7.2 Open-Circuit Time Constants

$T'_{do}$, $T''_{do}$, $T'_{qo}$, and $T''_{qo}$ are called open-circuit time constants since these are the time constants under open-circuit condition. Suppose the synchronous generator is run at $\omega = \omega_o$, and then, the switch shown in Fig. 1.16 is closed. During this condition, $i_d = i_q = i_0 = 0$. The equations governing the synchronous generator during this condition are

$$\frac{d\psi_F}{dt} = \frac{1}{T'_d}\left[-\psi_F + \psi_d + \frac{L'_d(T''_{dc} - T'_d)}{(L'_d - L_d)(T'_d - T''_d)}E_f\right] \qquad (1.217)$$

$$\frac{d\psi_{1D}}{dt} = \frac{1}{T''_d}\left[-\psi_{1D} + \psi_d + \frac{L'_d L''_d(T''_{dc} - T'_d)}{L_d(L''_d - L'_d)(T''_d - T'_d)}E_f\right] \qquad (1.218)$$

$$\frac{d\psi_{1Q}}{dt} = \frac{1}{T'_q}\left(-\psi_{1Q} + \psi_q\right) \qquad (1.219)$$

$$\frac{d\psi_{2Q}}{dt} = \frac{1}{T''_q}\left(-\psi_{2Q} + \psi_q\right) \qquad (1.220)$$

$$0 = \frac{1}{L''_d}\psi_d + \left(\frac{1}{L_d} - \frac{1}{L'_d}\right)\psi_F + \left(\frac{1}{L'_d} - \frac{1}{L''_d}\right)\psi_{1D} \qquad (1.221)$$

$$0 = \frac{1}{L''_q}\psi_q + \left(\frac{1}{L_q} - \frac{1}{L'_q}\right)\psi_{1Q} + \left(\frac{1}{L'_q} - \frac{1}{L''_q}\right)\psi_{2Q} \qquad (1.222)$$

The last two equations imply that $\psi_d$ and $\psi_q$ are not independent and can be expressed in terms of state variables. Using (1.191), (1.193), (1.197), and (1.198), it can be shown that the eigenvalues of the system during this condition are $-1/T'_{do}, -1/T''_{do}, -1/T'_{qo}, -1/T''_{qo}$. Hence, $T'_{do}$, $T''_{do}$, $T'_{qo}$, and $T''_{qo}$ are called open-circuit time constants.

## 1.8  Model in Per Unit Quantities

It is convenient to analyze the equations governing a power system, if the values of
the quantities (variables and parameters) are normalized by dividing them by their
respective base values. The normalized value is said to be in per unit of the base
value. The base values of some quantities can be chosen independently, and from
these values, the base values of other quantities are obtained. If the base values of
angular frequency, power, and voltage are chosen, the base values of other quantities
can be obtained. Base angular frequency and base power are same for the entire
power system, whereas base voltage is different on the two sides of a transformer.
The obvious choice for base angular frequency $\omega_B$ is the rated or nominal angular
frequency. A choice for the base power $S_B$ is the rated voltamperes of the largest
synchronous generator in the system. Base voltage $V_B$ is taken as the rated rms value
of line-to-line voltage. The base values of other quantities are obtained as follows:

$$\text{Base current,} \quad I_B \triangleq \frac{S_B}{V_B} \tag{1.223}$$

$$\text{Base impedance,} \quad Z_B \triangleq \frac{V_B}{I_B} \tag{1.224}$$

$$\text{Base flux linkage,} \quad \psi_B \triangleq \frac{V_B}{\omega_B} \tag{1.225}$$

$$\text{Base inductance,} \quad L_B \triangleq \frac{\psi_B}{I_B} \tag{1.226}$$

$$\text{Base torque,} \quad T_B \triangleq \frac{S_B}{\omega_B} \tag{1.227}$$

The quantities in per unit are denoted by a bar over the notation; for example,
$\bar{\psi}_d \triangleq \psi_d/\psi_B$. Dividing (1.199)–(1.205) by $\psi_B$ gives

$$\frac{d\bar{\psi}_d}{dt} = -\omega\bar{\psi}_q - \omega_B \bar{R}_a \bar{i}_d - \omega_B \bar{v}_d \tag{1.228}$$

$$\frac{d\bar{\psi}_q}{dt} = \omega\bar{\psi}_d - \omega_B \bar{R}_a \bar{i}_q - \omega_B \bar{v}_q \tag{1.229}$$

$$\frac{d\bar{\psi}_0}{dt} = -\omega_B \bar{R}_a \bar{i}_0 - \omega_B \bar{v}_0 \tag{1.230}$$

$$\frac{d\bar{\psi}_F}{dt} = \frac{1}{T_d'}\left[-\bar{\psi}_F + \bar{\psi}_d + \frac{\bar{X}_d'(T_{dc}'' - T_d')}{(\bar{X}_d' - \bar{X}_d)(T_d' - T_d'')}\bar{E}_f\right] \tag{1.231}$$

$$\frac{d\bar{\psi}_{1D}}{dt} = \frac{1}{T_d''}\left[-\bar{\psi}_{1D} + \bar{\psi}_d + \frac{\bar{X}_d'\bar{X}_d''(T_{dc}'' - T_d'')}{\bar{X}_d(\bar{X}_d'' - \bar{X}_d')(T_d'' - T_d')}\bar{E}_f\right] \tag{1.232}$$

$$\frac{d\bar{\psi}_{1Q}}{dt} = \frac{1}{T_q'}\left(-\bar{\psi}_{1Q} + \bar{\psi}_q\right) \tag{1.233}$$

$$\frac{d\bar{\psi}_{2Q}}{dt} = \frac{1}{T_q''}\left(-\bar{\psi}_{2Q} + \bar{\psi}_q\right) \tag{1.234}$$

where $X_d \triangleq \omega_B L_d$, $X_d' \triangleq \omega_B L_d'$, and $X_d'' \triangleq \omega_B L_d''$. It is to be noted that $\bar{X}_d = \bar{L}_d$, $\bar{X}_d' = \bar{L}_d'$, and $\bar{X}_d'' = \bar{L}_d''$.

Equation (1.207) can be written as

$$\frac{d\omega}{dt} = \frac{S_B}{J'\omega_B}\left(\frac{T_m'}{T_B} - \frac{\psi_d}{\psi_B}\frac{i_q}{I_B} + \frac{\psi_q}{\psi_B}\frac{i_d}{I_B}\right) \tag{1.235}$$

A parameter called inertia constant denoted by $H$ is defined [1–3] as

$$H \triangleq \frac{J'\omega_B^2}{2S_B} \tag{1.236}$$

Equation (1.235) can be written as

$$\frac{d\omega}{dt} = \frac{\omega_B}{2H}(\bar{T}_m' - \bar{\psi}_d\bar{i}_q + \bar{\psi}_q\bar{i}_d) \tag{1.237}$$

Dividing (1.208)–(1.210) by $I_B$ gives

$$\bar{i}_d = \frac{1}{\bar{X}_d''}\bar{\psi}_d + \left(\frac{1}{\bar{X}_d} - \frac{1}{\bar{X}_d'}\right)\bar{\psi}_F + \left(\frac{1}{\bar{X}_d'} - \frac{1}{\bar{X}_d''}\right)\bar{\psi}_{1D} \tag{1.238}$$

$$\bar{i}_q = \frac{1}{\bar{X}_q''}\bar{\psi}_q + \left(\frac{1}{\bar{X}_q} - \frac{1}{\bar{X}_q'}\right)\bar{\psi}_{1Q} + \left(\frac{1}{\bar{X}_q'} - \frac{1}{\bar{X}_q''}\right)\bar{\psi}_{2Q} \tag{1.239}$$

$$\bar{i}_0 = \frac{1}{\bar{X}_0}\bar{\psi}_0 \tag{1.240}$$

where $X_q \triangleq \omega_B L_q$, $X_q' \triangleq \omega_B L_q'$, $X_q'' \triangleq \omega_B L_q''$, and $X_0 = \omega_B L_0$. It is to be noted that $\bar{X}_q = \bar{L}_q$, $\bar{X}_q' = \bar{L}_q'$, $\bar{X}_q'' = \bar{L}_q''$, and $\bar{X}_0 = \bar{L}_0$. Equations (1.206), (1.228)–(1.234), and (1.237)–(1.240) form the synchronous generator model in per unit quantities.

## 1.9  Other Models

The equations obtained in the last section give a detailed model of the synchronous generator. The choice of the model depends on the application. Some applications do not need a detailed model. The simpler models in the order of decreasing complexity are given below.

### 1.9.1 Model in the Absence of Zero Sequence Variables and with $T''_{dc} = T''_d$

If the zero sequence variables are equal to zero and if it is assumed that $T''_{dc} = T''_d$, the equations governing the synchronous generator are

$$\frac{d\bar{\psi}_d}{dt} = -\omega\bar{\psi}_q - \omega_B \bar{R}_a \bar{i}_d - \omega_B \bar{v}_d \tag{1.241}$$

$$\frac{d\bar{\psi}_q}{dt} = \omega\bar{\psi}_d - \omega_B \bar{R}_a \bar{i}_q - \omega_B \bar{v}_q \tag{1.242}$$

$$\frac{d\bar{\psi}_F}{dt} = \frac{1}{T'_d}\left(-\bar{\psi}_F + \bar{\psi}_d + \frac{\bar{X}'_d}{\bar{X}_d - \bar{X}'_d}\bar{E}_f\right) \tag{1.243}$$

$$\frac{d\bar{\psi}_{1D}}{dt} = \frac{1}{T''_d}\left(-\bar{\psi}_{1D} + \bar{\psi}_d\right) \tag{1.244}$$

$$\frac{d\bar{\psi}_{1Q}}{dt} = \frac{1}{T'_q}\left(-\bar{\psi}_{1Q} + \bar{\psi}_q\right) \tag{1.245}$$

$$\frac{d\bar{\psi}_{2Q}}{dt} = \frac{1}{T''_q}\left(-\bar{\psi}_{2Q} + \bar{\psi}_q\right) \tag{1.246}$$

$$\frac{d\delta}{dt} = \omega - \omega_o \tag{1.247}$$

$$\frac{d\omega}{dt} = \frac{\omega_B}{2H}(\bar{T}'_m - \bar{\psi}_d \bar{i}_q + \bar{\psi}_q \bar{i}_d) \tag{1.248}$$

$$\bar{i}_d = \frac{1}{\bar{X}''_d}\bar{\psi}_d + \left(\frac{1}{\bar{X}_d} - \frac{1}{\bar{X}'_d}\right)\bar{\psi}_F + \left(\frac{1}{\bar{X}'_d} - \frac{1}{\bar{X}''_d}\right)\bar{\psi}_{1D} \tag{1.249}$$

$$\bar{i}_q = \frac{1}{\bar{X}''_q}\bar{\psi}_q + \left(\frac{1}{\bar{X}_q} - \frac{1}{\bar{X}'_q}\right)\bar{\psi}_{1Q} + \left(\frac{1}{\bar{X}'_q} - \frac{1}{\bar{X}''_q}\right)\bar{\psi}_{2Q} \tag{1.250}$$

### 1.9.2 Model with Stator Transients Neglected

If high-frequency transients are not of interest, stator transients are neglected by neglecting the terms $d\bar{\psi}_d/dt$ and $d\bar{\psi}_q/dt$ in (1.241) and (1.242), respectively. Then, these equations become algebraic equations.

$$0 = -\omega\bar{\psi}_q - \omega_B \bar{R}_a \bar{i}_d - \omega_B \bar{v}_d \tag{1.251}$$

$$0 = \omega\bar{\psi}_d - \omega_B \bar{R}_a \bar{i}_q - \omega_B \bar{v}_q \tag{1.252}$$

Another assumption made is that $\omega = \omega_B$ in (1.251) and (1.252). The synchronous generator model is given by the following equations:

$$\frac{d\bar{\psi}_F}{dt} = \frac{1}{T_d'}\left(-\bar{\psi}_F + \bar{\psi}_d + \frac{\bar{X}_d'}{\bar{X}_d - \bar{X}_d'}\bar{E}_f\right) \tag{1.253}$$

$$\frac{d\bar{\psi}_{1D}}{dt} = \frac{1}{T_d''}\left(-\bar{\psi}_{1D} + \bar{\psi}_d\right) \tag{1.254}$$

$$\frac{d\bar{\psi}_{1Q}}{dt} = \frac{1}{T_q'}\left(-\bar{\psi}_{1Q} + \bar{\psi}_q\right) \tag{1.255}$$

$$\frac{d\bar{\psi}_{2Q}}{dt} = \frac{1}{T_q''}\left(-\bar{\psi}_{2Q} + \bar{\psi}_q\right) \tag{1.256}$$

$$\frac{d\delta}{dt} = \omega - \omega_o \tag{1.257}$$

$$\frac{d\omega}{dt} = \frac{\omega_B}{2H}(\bar{T}_m' - \bar{\psi}_d \bar{i}_q + \bar{\psi}_q \bar{i}_d) \tag{1.258}$$

$$0 = -\bar{\psi}_q - \bar{R}_a \bar{i}_d - \bar{v}_d \tag{1.259}$$

$$0 = \bar{\psi}_d - \bar{R}_a \bar{i}_q - \bar{v}_q \tag{1.260}$$

$$\bar{i}_d = \frac{1}{\bar{X}_d''}\bar{\psi}_d + \left(\frac{1}{\bar{X}_d} - \frac{1}{\bar{X}_d'}\right)\bar{\psi}_F + \left(\frac{1}{\bar{X}_d'} - \frac{1}{\bar{X}_d''}\right)\bar{\psi}_{1D} \tag{1.261}$$

$$\bar{i}_q = \frac{1}{\bar{X}_q''}\bar{\psi}_q + \left(\frac{1}{\bar{X}_q} - \frac{1}{\bar{X}_q'}\right)\bar{\psi}_{1Q} + \left(\frac{1}{\bar{X}_q'} - \frac{1}{\bar{X}_q''}\right)\bar{\psi}_{2Q} \tag{1.262}$$

### 1.9.3  Two Axis Model

The time constants $T_d''$ and $T_q''$ are set to zero, and hence,

$$\bar{\psi}_{1D} = \bar{\psi}_d \tag{1.263}$$

$$\bar{\psi}_{2Q} = \bar{\psi}_q \tag{1.264}$$

The synchronous generator model is given by the following equations:

$$\frac{d\bar{\psi}_F}{dt} = \frac{1}{T_d'}\left(-\bar{\psi}_F + \bar{\psi}_d + \frac{\bar{X}_d'}{\bar{X}_d - \bar{X}_d'}\bar{E}_f\right) \tag{1.265}$$

$$\frac{d\bar{\psi}_{1Q}}{dt} = \frac{1}{T_q'}\left(-\bar{\psi}_{1Q} + \bar{\psi}_q\right) \tag{1.266}$$

$$\frac{d\delta}{dt} = \omega - \omega_o \tag{1.267}$$

$$\frac{d\omega}{dt} = \frac{\omega_B}{2H}(\bar{T}'_m - \bar{\psi}_d \bar{i}_q + \bar{\psi}_q \bar{i}_d) \tag{1.268}$$

$$0 = -\bar{\psi}_q - \bar{R}_a \bar{i}_d - \bar{v}_d \tag{1.269}$$

$$0 = \bar{\psi}_d - \bar{R}_a \bar{i}_q - \bar{v}_q \tag{1.270}$$

$$\bar{i}_d = \frac{1}{\bar{X}'_d}\bar{\psi}_d + \left(\frac{1}{\bar{X}_d} - \frac{1}{\bar{X}'_d}\right)\bar{\psi}_F \tag{1.271}$$

$$\bar{i}_q = \frac{1}{\bar{X}'_q}\bar{\psi}_q + \left(\frac{1}{\bar{X}_q} - \frac{1}{\bar{X}'_q}\right)\bar{\psi}_{1Q} \tag{1.272}$$

### 1.9.4 One Axis (Flux Decay) Model

The time constant $T'_q$ is set to zero, and hence,

$$\bar{\psi}_{1Q} = \bar{\psi}_q \tag{1.273}$$

The synchronous generator model is given by the following equations:

$$\frac{d\bar{\psi}_F}{dt} = \frac{1}{T'_d}\left(-\bar{\psi}_F + \bar{\psi}_d + \frac{\bar{X}'_d}{\bar{X}_d - \bar{X}'_d}\bar{E}_f\right) \tag{1.274}$$

$$\frac{d\delta}{dt} = \omega - \omega_o \tag{1.275}$$

$$\frac{d\omega}{dt} = \frac{\omega_B}{2H}(\bar{T}'_m - \bar{\psi}_d \bar{i}_q + \bar{\psi}_q \bar{i}_d) \tag{1.276}$$

$$0 = -\bar{\psi}_q - \bar{R}_a \bar{i}_d - \bar{v}_d \tag{1.277}$$

$$0 = \bar{\psi}_d - \bar{R}_a \bar{i}_q - \bar{v}_q \tag{1.278}$$

$$\bar{i}_d = \frac{1}{\bar{X}'_d}\bar{\psi}_d + \left(\frac{1}{\bar{X}_d} - \frac{1}{\bar{X}'_d}\right)\bar{\psi}_F \tag{1.279}$$

$$\bar{i}_q = \frac{1}{\bar{X}_q}\bar{\psi}_q \tag{1.280}$$

### *1.9.5 Classical Model*

$T'_d$ is assumed to be infinite. This implies $d\bar{\psi}_F/dt = 0$, i.e., $\bar{\psi}_F$ is constant. It is also assumed that $\bar{R}_a = 0$ and $\bar{X}_q = \bar{X}'_d$. Since $\bar{R}_a = 0$, $\bar{\psi}_q = -\bar{v}_d$ and $\bar{\psi}_d = \bar{v}_q$. The synchronous generator model is given by the following equations:

$$\frac{d\delta}{dt} = \omega - \omega_o \tag{1.281}$$

$$\frac{d\omega}{dt} = \frac{\omega_B}{2H}(\bar{T}'_m - \bar{v}_q\bar{i}_q - \bar{v}_d\bar{i}_d) \tag{1.282}$$

$$\bar{i}_d = \frac{1}{\bar{X}'_d}\bar{v}_q - \frac{\bar{E}}{\bar{X}'_d} \tag{1.283}$$

$$\bar{i}_q = -\frac{1}{\bar{X}'_d}\bar{v}_d \tag{1.284}$$

where $\bar{E} \triangleq (\bar{X}_d - \bar{X}'_d)\bar{\psi}_F/\bar{X}_d$ is a constant.

## References

1. K.R. Padiyar, *Power System Dynamics: Stability and Control*, 2nd edn. (BS Publications, Hyderabad, 2002)
2. P. Kundur, *Power System Stability and Control* (Tata McGraw-Hill, Noida, 1994)
3. P.W. Sauer, M.A. Pai, *Power System Dynamics and Stability* (Pearson Education, Singapore, 1998)
4. P.C. Krause, *Analysis of Electric Machinery* (McGraw-Hill, New York, 1987)
5. P.M. Anderson, B.L. Agrawal, J.E. Van Ness, *Subsynchronous Resonance in Power Systems* (IEEE Press, New York, 1990)
6. A.M. Kulkarni, Power system dynamics and control (2012), http://www.nptel.iitm.ac.in/courses/108101004/

# Chapter 2
# Transformer, Transmission Line, and Load

## 2.1 Transformer

### 2.1.1 Single-Phase Transformer

The single-phase transformer consists of a core and two or more windings. Figure 2.1 shows a transformer with two windings.

Let the transformer be ideal: the windings have zero resistance and the core has infinite permeability [1]. Infinite permeability means that there is no flux outside the core.

$N_1$ and $N_2$ are the number of turns in the windings. If the flux in the core is $\phi$, the induced emfs in the windings are

$$e_1 = N_1 \frac{d\phi}{dt} \tag{2.1}$$

$$e_2 = N_2 \frac{d\phi}{dt} \tag{2.2}$$

From (2.1) and (2.2),

$$\frac{e_1}{e_2} = \frac{N_1}{N_2} \tag{2.3}$$

The relation between $i_1$ and $i_2$ is obtained from Ampere's law. Due to infinite permeability, the magnetic field intensity in the core is zero. Application of Ampere's law to the closed path in the core, shown in Fig. 2.1, gives

$$\frac{i_1}{i_2} = \frac{N_2}{N_1} \tag{2.4}$$

S Krishna, *An Introduction to Modelling of Power System Components*, 
SpringerBriefs in Electrical and Computer Engineering, 
DOI: 10.1007/978-81-322-1847-0_2, © The Author(s) 2014

**Fig. 2.1** Single-phase
transformer with two windings

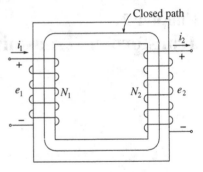

**Fig. 2.2** Single-phase
transformer with three
windings

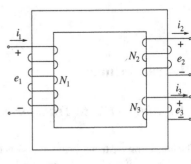

**Fig. 2.3** Representation of
ideal transformer

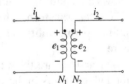

For the three-winding transformer shown in Fig. 2.2, where the number of turns
in the windings are $N_1$, $N_2$, and $N_3$,

$$\frac{e_1}{e_2} = \frac{N_1}{N_2}, \frac{e_1}{e_3} = \frac{N_1}{N_3} \tag{2.5}$$

$$N_1 i_1 = N_2 i_2 + N_3 i_3 \tag{2.6}$$

The two-winding ideal transformer can be represented by the equivalent circuit
shown in Fig. 2.3. The dots shown at a terminal of each winding indicate the winding
terminals which simultaneously have the same polarity due to the emfs induced.

There are applications where the ideal transformer cannot be used. Then the
equivalent circuit of the transformer is given by Fig. 2.4. $R_1$ and $R_2$ are the resistances
of the two windings. Though the permeability of the core is high, it is not infinite,
and hence there is flux outside the core which links some or all turns of only one
winding and induces an emf. This flux is called leakage flux and its effect is modelled
by leakage inductances $L_1$ and $L_2$. $e_1$ and $e_2$ are related by (2.3). Due to finite
permeability of the core, (2.4) is not exact but is used as an approximation.

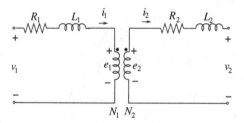

**Fig. 2.4** Equivalent circuit of transformer

## 2.1.2 Three-Phase Transformer

A three-phase transformer can be obtained from three identical single-phase transformers. Figure 2.5 shows the equivalent circuit of the wye-wye-connected transformer. The two windings of a single-phase transformer are shown parallel to each other.

Let $v_{1a}$, $v_{1b}$, and $v_{1c}$ be the potentials of the terminals $1a$, $1b$, and $1c$, respectively, with respect to the neutral. Let $v_{2a}$, $v_{2b}$, and $v_{2c}$ be the potentials of the terminals $2a$, $2b$, and $2c$, respectively, with respect to the neutral. The equations for the wye-wye-connected transformer are

$$v_{1a} = i_{1a}R_1 + L_1\frac{di_{1a}}{dt} + e_{1a} \tag{2.7}$$

$$v_{1b} = i_{1b}R_1 + L_1\frac{di_{1b}}{dt} + e_{1b} \tag{2.8}$$

$$v_{1c} = i_{1c}R_1 + L_1\frac{di_{1c}}{dt} + e_{1c} \tag{2.9}$$

$$v_{2a} = \frac{N_2}{N_1}e_{1a} - \frac{N_1}{N_2}i_{1a}R_2 - \frac{N_1}{N_2}L_2\frac{di_{1a}}{dt} \tag{2.10}$$

$$v_{2b} = \frac{N_2}{N_1}e_{1b} - \frac{N_1}{N_2}i_{1b}R_2 - \frac{N_1}{N_2}L_2\frac{di_{1b}}{dt} \tag{2.11}$$

$$v_{2c} = \frac{N_2}{N_1}e_{1c} - \frac{N_1}{N_2}i_{1c}R_2 - \frac{N_1}{N_2}L_2\frac{di_{1c}}{dt} \tag{2.12}$$

Elimination of induced emfs from (2.7) to (2.12) gives

$$v_{2a} = \frac{N_2}{N_1}v_{1a} - \left(\frac{N_2}{N_1}R_1 + \frac{N_1}{N_2}R_2\right)i_{1a} - \left(\frac{N_2}{N_1}L_1 + \frac{N_1}{N_2}L_2\right)\frac{di_{1a}}{dt} \tag{2.13}$$

$$v_{2b} = \frac{N_2}{N_1}v_{1b} - \left(\frac{N_2}{N_1}R_1 + \frac{N_1}{N_2}R_2\right)i_{1b} - \left(\frac{N_2}{N_1}L_1 + \frac{N_1}{N_2}L_2\right)\frac{di_{1b}}{dt} \tag{2.14}$$

$$v_{2c} = \frac{N_2}{N_1}v_{1c} - \left(\frac{N_2}{N_1}R_1 + \frac{N_1}{N_2}R_2\right)i_{1c} - \left(\frac{N_2}{N_1}L_1 + \frac{N_1}{N_2}L_2\right)\frac{di_{1c}}{dt} \tag{2.15}$$

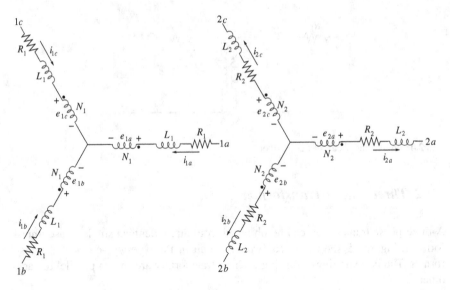

**Fig. 2.5** Wye-wye-connected transformer

In order to obtain the equations in per unit quantities, the equations are divided by the base voltage. If $V_{1B}$ is the base voltage on the transformer side with $N_1$ turns, the base voltage on the other side of the transformer is

$$V_{2B} \triangleq \frac{N_2}{N_1} V_{1B} \tag{2.16}$$

The base values of other quantities are obtained as follows.

$$I_{1B} = \frac{S_B}{V_{1B}}, \ I_{2B} = \frac{S_B}{V_{2B}}, \ Z_{1B} = \frac{V_{1B}}{I_{1B}}, \ Z_{2B} = \frac{V_{2B}}{I_{2B}} \tag{2.17}$$

Dividing (2.13)–(2.15) by $V_{2B}$ gives

$$\bar{v}_{2a} = \bar{v}_{1a} - \left(\overline{R}_1 + \overline{R}_2\right)\bar{i}_a - \frac{1}{\omega_B}\left(\overline{X}_1 + \overline{X}_2\right)\frac{d\bar{i}_a}{dt} \tag{2.18}$$

$$\bar{v}_{2b} = \bar{v}_{1b} - \left(\overline{R}_1 + \overline{R}_2\right)\bar{i}_b - \frac{1}{\omega_B}\left(\overline{X}_1 + \overline{X}_2\right)\frac{d\bar{i}_b}{dt} \tag{2.19}$$

$$\bar{v}_{2c} = \bar{v}_{1c} - \left(\overline{R}_1 + \overline{R}_2\right)\bar{i}_c - \frac{1}{\omega_B}\left(\overline{X}_1 + \overline{X}_2\right)\frac{d\bar{i}_c}{dt} \tag{2.20}$$

where $X_1 \triangleq \omega_B L_1$ and $X_2 \triangleq \omega_B L_2$. The subscripts 1 and 2 are not necessary in the notation for currents since $\bar{i}_{1a} = \bar{i}_{2a}, \bar{i}_{1b} = \bar{i}_{2b}$, and $\bar{i}_{1c} = \bar{i}_{2c}$.

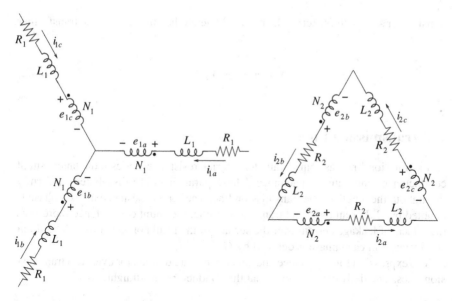

**Fig. 2.6** Wye-delta-connected transformer

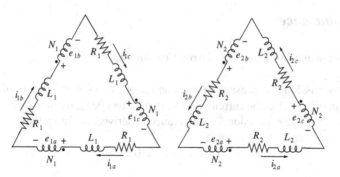

**Fig. 2.7** Delta-delta-connected transformer

The equivalent circuit of the wye-delta-connected transformer is shown in Fig. 2.6. For the wye-delta-connected transformer, if $V_{1B}$ is the base voltage on the transformer side with $N_1$ turns, the base voltage on the other side of the transformer is

$$V_{2B} \triangleq \frac{N_2}{\sqrt{3}N_1} V_{1B} \tag{2.21}$$

For balanced sinusoidal operation, if resistance and leakage inductance are neglected, the phase shift between terminal voltages on the two sides of the transformer is 30°.

The equivalent circuit of the delta-delta-connected transformer is shown in Fig. 2.7. For the delta-delta-connected transformer, if $V_{1B}$ is the base voltage on the transformer side with $N_1$ turns, the base voltage on the other side of the transformer is

$$V_{2B} \triangleq \frac{N_2}{N_1} V_{1B} \qquad (2.22)$$

.

## 2.2 Transmission Line

A transmission line has four parameters: series resistance, series inductance, shunt conductance, and shunt capacitance. These parameters are distributed uniformly throughout the length of the transmission line. The series resistance in each phase is denoted by $R$. For an overhead transmission line, the shunt conductance represents the effects of leakage current over the surface of the insulator and corona. The shunt conductance in each phase is denoted by $G$.

The expression for inductance and capacitance are derived for overhead transmission lines. The derivations assume that the conductors are straight.

### 2.2.1 Inductance

#### 2.2.1.1 Transmission Line with Three Conductors

Let the transmission line consist of three conductors, one for each phase, of radius $r$ as shown in Fig. 2.8. Let the current in these conductors be $i_a$, $i_b$, and $i_c$ with uniform current density. The expression for inductance is derived assuming that

$$i_a + i_b + i_c = 0 \qquad (2.23)$$

It is assumed that the three conductors are transposed if not spaced symmetrically, in order to have a symmetrical system; the transmission line is divided into three sections of equal lengths and each conductor occupies each of the three positions 1, 2, and 3 for one third of the transmission line length. Let the conductors $a, b, c$ occupy positions 1, 2, 3, respectively, in the first section, positions 2, 3, 1, respectively, in the second section, and positions 3, 1, 2, respectively, in the third section.

Consider a tube of radius $x < r$ and thickness $dx$ in phase $a$ conductor in section 1 as shown in Fig. 2.8; the tube is coaxial with the conductor. Consider a filament in this tube with cross-sectional area $x dx d\theta$; $d\theta$ is the angle subtended at the axis of the conductor by the filament [2]. Consider the closed path consisting of this filament and an arbitrarily located (at P) straight line parallel to the conductors. Let $\psi_{fa}$, $\psi_{fb}$, and $\psi_{fc}$ be the flux linkage of this closed path in section 1, due to $i_a$, $i_b$, and $i_c$, respectively. The power delivered to this closed path, due to $i_a$, is equal to

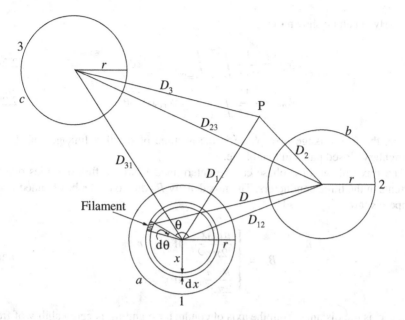

**Fig. 2.8** Cross section of transmission line conductors

$$\frac{d\psi_{fa}}{dt} i_a \frac{x dx d\theta}{\pi r^2} = \frac{1}{\pi r^2} \psi_{fa} \frac{di_a}{dt} x dx d\theta \tag{2.24}$$

The power delivered to phase $a$, due to $i_a$, is obtained by integrating this expression over the cross-sectional area of the conductor as follows.

$$p = \frac{1}{\pi r^2} \frac{di_a}{dt} \int_{\theta=0}^{2\pi} \int_{x=0}^{r} \psi_{fa} x dx d\theta \tag{2.25}$$

Let $\psi_{aa}$, $\psi_{ab}$, and $\psi_{ac}$ be the flux linkages of phase $a$ in section 1 due to $i_a$, $i_b$, and $i_c$, respectively. The expression for $p$ can also be written in terms of flux linkage of phase $a$ due to $i_a$, as

$$p = \frac{d\psi_{aa}}{dt} i_a = \psi_{aa} \frac{di_a}{dt} \tag{2.26}$$

From (2.25) and (2.26),

$$\psi_{aa} = \frac{1}{\pi r^2} \int_{\theta=0}^{2\pi} \int_{x=0}^{r} \psi_{fa} x dx d\theta \tag{2.27}$$

Similarly, it can be shown that

$$\psi_{ab} = \frac{1}{\pi r^2} \int_{\theta=0}^{2\pi} \int_{x=0}^{r} \psi_{fb} x \, dx \, d\theta \tag{2.28}$$

$$\psi_{ac} = \frac{1}{\pi r^2} \int_{\theta=0}^{2\pi} \int_{x=0}^{r} \psi_{fc} x \, dx \, d\theta \tag{2.29}$$

Hence, the flux linkage of a phase is the average of the flux linkages of all the filamentary closed paths in that phase.

The flux linkage of a phase can be determined from the flux densities due to currents in the three conductors. The flux density $B_a$ due to $i_a$ can be obtained from Ampere's law.

$$B_a = \begin{cases} \dfrac{\mu_0 i_a x'}{2\pi r^2} & \text{if } x' \le r \\[2ex] \dfrac{\mu_0 i_a}{2\pi x'} & \text{if } x' \ge r \end{cases} \tag{2.30}$$

where $x'$ is the distance from the axis of conductor $a$ and $\mu_0$ is permeability of free space; permeability of air and conductor are almost equal to that of free space. Then,

$$\psi_{fa} = \frac{l}{3} \int_{x}^{D_1} B_a dx' = \frac{\mu_0 i_a l}{6\pi} \left( \frac{1}{2} - \frac{x^2}{2r^2} + \ln \frac{D_1}{r} \right) \tag{2.31}$$

where $l$ is the length of the transmission line. From (2.27) and (2.31),

$$\psi_{aa} = \frac{1}{\pi r^2} \int_{\theta=0}^{2\pi} \int_{x=0}^{r} \frac{\mu_0 i_a l}{6\pi} \left( \frac{1}{2} - \frac{x^2}{2r^2} + \ln \frac{D_1}{r} \right) x \, dx \, d\theta = \frac{\mu_0 i_a l}{6\pi} \ln \frac{D_1}{r'} \tag{2.32}$$

where $r' \triangleq e^{-1/4} r$.

The flux density due to $i_b$ is

$$B_b = \frac{\mu_0 i_b}{2\pi D'} \quad \text{if } D' \ge r \tag{2.33}$$

where $D'$ is the distance from the axis of conductor $b$. Then,

$$\psi_{fb} = \frac{l}{3} \int_{D}^{D_2} B_b dD' = \frac{\mu_0 i_b l}{6\pi} \ln \frac{D_2}{D} \tag{2.34}$$

where $D = \left( D_{12}^2 + x^2 - 2D_{12} x \cos \theta \right)^{1/2}$. From (2.28) and (2.34),

$$\psi_{ab} = \frac{1}{\pi r^2} \int_{\theta=0}^{2\pi} \int_{x=0}^{r} \frac{\mu_0 i_b l}{6\pi} \ln \frac{D_2}{D} x \, dx \, d\theta = \frac{\mu_0 i_b l}{6\pi} \ln \frac{D_2}{D_{12}} \tag{2.35}$$

Similarly,

$$\psi_{ac} = \frac{\mu_0 i_c l}{6\pi} \ln \frac{D_3}{D_{31}} \tag{2.36}$$

The flux linkage of phase $a$ in section 1 is

$$\psi_{a1} = \psi_{aa} + \psi_{ab} + \psi_{ac} = \frac{\mu_0 l}{6\pi} \left( i_a \ln \frac{D_1}{r'} + i_b \ln \frac{D_2}{D_{12}} + i_c \ln \frac{D_3}{D_{31}} \right) \tag{2.37}$$

Similarly, the flux linkage of phase $a$ in sections 2 and 3, $\psi_{a2}$ and $\psi_{a3}$, respectively, are given by

$$\psi_{a2} = \frac{\mu_0 l}{6\pi} \left( i_a \ln \frac{D_2}{r'} + i_b \ln \frac{D_3}{D_{23}} + i_c \ln \frac{D_1}{D_{12}} \right) \tag{2.38}$$

$$\psi_{a3} = \frac{\mu_0 l}{6\pi} \left( i_a \ln \frac{D_3}{r'} + i_b \ln \frac{D_1}{D_{31}} + i_c \ln \frac{D_2}{D_{23}} \right) \tag{2.39}$$

The flux linkage of phase $a$ is

$$\psi_a = \psi_{a1} + \psi_{a2} + \psi_{a3} \tag{2.40}$$

From (2.37) to (2.40),

$$\psi_a = \frac{\mu_0 l}{2\pi} \left[ i_a \ln \frac{(D_1 D_2 D_3)^{1/3}}{r'} + i_b \ln \left( \frac{D_1 D_2 D_3}{D_{12} D_{23} D_{31}} \right)^{1/3} + i_c \ln \left( \frac{D_1 D_2 D_3}{D_{12} D_{23} D_{31}} \right)^{1/3} \right] \tag{2.41}$$

The coefficient of $i_a$ is self inductance and the coefficients of $i_b$ and $i_c$ are mutual inductances. Using (2.23), the self and mutual inductances can be replaced by an equivalent self inductance $L$.

$$L = \frac{\mu_0 l}{2\pi} \ln \frac{(D_{12} D_{23} D_{31})^{1/3}}{r'} \tag{2.42}$$

### 2.2.1.2 Composite Conductors

A composite conductor consists of two or more individual conductors. Examples of composite conductor are bundled conductor, stranded conductor, and conductor of a multi-circuit transmission line. Figures 2.9, 2.10, and 2.11 show a double circuit transmission line, a transmission line with bundled conductors, and a stranded conductor, respectively.

**Fig. 2.9** Double circuit
transmission line

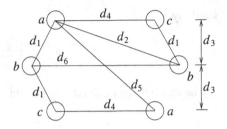

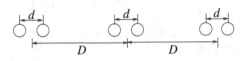

**Fig. 2.10** Transmission line
with bundled conductors

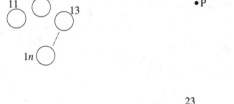

**Fig. 2.11** Stranded conductor

**Fig. 2.12** Transmission line
with composite conductors

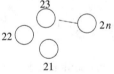

Consider the transmission line consisting of a composite conductor in each phase, as shown in Fig. 2.12 [2]. Let each phase consist of $n$ individual conductors of radius $r$.

It is assumed that the three phases are transposed if not placed symmetrically; the transmission line is divided into three sections of equal lengths and each phase occupies each of the three positions 1, 2, and 3 for one third of the transmission line length. Let phases $a, b, c$ occupy positions 1, 2, 3, respectively, in the first section, positions 2, 3, 1, respectively, in the second section, and positions 3, 1, 2, respectively, in the third section. The position of each individual conductor is identified by two numbers as in Fig. 2.12; the first number is that of the position of the phase and the second number is that of the position of the individual conductor. Let the current in the individual conductors of phases $a, b, c$ be $i_a/n, i_b/n, i_c/n$, respectively. This is true if the individual conductors in each phase are transposed so that each individual conductor occupies each of the $n$ positions for equal lengths along a section. $i_a$, $i_b$, and $i_c$ satisfy (2.23). Consider the closed path consisting of the individual conductor of phase $a$ at position $1k$ for length $l/(3n)$, and the straight line (at P) parallel to the conductors. Similar to (2.37), the flux linkage of this closed path is

$$\psi_{a1k} = \frac{\mu_0 l}{6\pi n^2} \sum_{m=1}^{n} \left[ i_a \ln \frac{D_{1m}}{D_{1k1m}} + i_b \ln \frac{D_{2m}}{D_{1k2m}} + i_c \ln \frac{D_{3m}}{D_{1k3m}} \right] \quad (2.43)$$

where $D_{pkqm}$ ($p$ and $q$ are 1, 2, or 3, and $pk \neq qm$) is the distance between the axes of conductors at positions $pk$ and $qm$, $D_{pkpk} = r'$, and $D_{pk}$ is the distance between point P and the axis of conductor at position $pk$. It is evident from (2.27) to (2.29) that the flux linkage of a phase is the average of the flux linkages of the closed paths formed by individual conductors in that phase. Therefore, the flux linkage of phase $a$ in section 1 is

$$\psi_{a1} = \frac{\mu_0 l}{6\pi n^2} \sum_{k=1}^{n} \sum_{m=1}^{n} \left[ i_a \ln \frac{D_{1m}}{D_{1k1m}} + i_b \ln \frac{D_{2m}}{D_{1k2m}} + i_c \ln \frac{D_{3m}}{D_{1k3m}} \right] \quad (2.44)$$

Similarly, the flux linkage of phase $a$ in sections 2 and 3, $\psi_{a2}$ and $\psi_{a3}$, respectively, are given by

$$\psi_{a2} = \frac{\mu_0 l}{6\pi n^2} \sum_{k=1}^{n} \sum_{m=1}^{n} \left[ i_a \ln \frac{D_{2m}}{D_{2k2m}} + i_b \ln \frac{D_{3m}}{D_{2k3m}} + i_c \ln \frac{D_{1m}}{D_{2k1m}} \right] \quad (2.45)$$

$$\psi_{a3} = \frac{\mu_0 l}{6\pi n^2} \sum_{k=1}^{n} \sum_{m=1}^{n} \left[ i_a \ln \frac{D_{3m}}{D_{3k3m}} + i_b \ln \frac{D_{1m}}{D_{3k1m}} + i_c \ln \frac{D_{2m}}{D_{3k2m}} \right] \quad (2.46)$$

The flux linkage of phase $a$ is

$$\psi_a = \psi_{a1} + \psi_{a2} + \psi_{a3} \quad (2.47)$$

From (2.23) and (2.44) to (2.47), the equivalent self inductance of each phase is

$$L = \frac{\mu_0 l}{2\pi} \ln \frac{D_m}{D_s} \quad (2.48)$$

where

$$D_m \triangleq \left( \prod_{k=1}^{n} \prod_{m=1}^{n} D_{1k2m} D_{2k3m} D_{3k1m} \right)^{1/(3n^2)} \quad (2.49)$$

$$D_s \triangleq \left( \prod_{k=1}^{n} \prod_{m=1}^{n} D_{1k1m} D_{2k2m} D_{3k3m} \right)^{1/(3n^2)} \quad (2.50)$$

$D_m$ is known as mutual geometric mean distance (GMD) and $D_s$ is known as self GMD.

For the double circuit transmission line shown in Fig. 2.9,

$$D_m = \left(2d_1^2 d_2^2 d_3 d_4\right)^{1/6} \tag{2.51}$$

$$D_s = \left(e^{-3/4} r^3 d_5^2 d_6\right)^{1/6} \tag{2.52}$$

where $r$ is the radius of the individual conductors. For hexagonal spacing ($d_4 = d_1$ and $d_6 = d_5$), transposition is not necessary.

For the transmission line with bundled conductors shown in Fig. 2.10, where each bundle (composite conductor) consists of two individual conductors,

$$D_m = \left[4D^6 \left(D^2 - d^2\right)^2 \left(4D^2 - d^2\right)\right]^{1/12} \tag{2.53}$$

$$D_s = \left(e^{-1/4} rd\right)^{1/2} \tag{2.54}$$

where $r$ is the radius of the individual conductors.

For the stranded conductor shown in Fig. 2.11,

$$D_s = 2\,(364.5)^{1/49}\, e^{-1/28} r \tag{2.55}$$

where $r$ is the radius of the strands.

It is to be noted that the transposition of the individual conductors in the composite conductor is not necessary for the three phases to be symmetrical; but the assumption of transposition helps in easily obtaining the expression for inductance.

### 2.2.2 Capacitance

#### 2.2.2.1 Transmission Line with Three Conductors

Consider the transmission line with three conductors shown in Fig. 2.8. Let the charge per unit length on the conductors of phases $a$, $b$, and $c$ be $q_a$, $q_b$, and $q_c$, respectively, such that

$$q_a + q_b + q_c = 0 \tag{2.56}$$

The radius of the conductors is assumed to be very small compared to the distance between any two conductors. Therefore, the potential of conductor $a$ with respect to the point P is

$$v_{aP} = \frac{1}{2\pi\varepsilon_0}\left(q_a \ln\frac{D_1}{r} + q_b \ln\frac{D_2}{D_{12}} + q_c \ln\frac{D_3}{D_{31}}\right) \tag{2.57}$$

$\varepsilon_0$ is the permittivity of free space; permittivity of air is almost equal to that of free space. The potential of the conductor is obtained by allowing P to recede to infinity. As P recedes to infinity, using (2.56), the potential of conductor $a$ for symmetrical spacing of conductors $(D_{12} = D_{23} = D_{31} = D)$ is

$$v_a = \frac{1}{2\pi\varepsilon_0}\left(q_a \ln\frac{1}{r} + q_b \ln\frac{1}{D} + q_c \ln\frac{1}{D}\right) \tag{2.58}$$

From (2.56) and (2.58), the capacitance in each phase is

$$C = \frac{2\pi\varepsilon_0 l}{\ln(D/r)} \tag{2.59}$$

If the conductors are not spaced symmetrically, it is assumed that transposition is done in order to have a symmetrical system. Let the conductors $a$, $b$, $c$ occupy positions 1, 2, 3, respectively, in the first section, positions 2, 3, 1, respectively, in the second section, and positions 3, 1, 2, respectively, in the third section. The charge per unit length is not same in all three sections for any phase whereas the potential is same. It is assumed that the charge per unit length is same in all the three sections [3]; let the charge per unit length on the conductors of phases $a$, $b$, and $c$ be $q_a$, $q_b$, and $q_c$, respectively, which satisfy (2.56). With this assumption, the potential of conductor $a$ in sections 1, 2, and 3, $v_{a1}$, $v_{a2}$, and $v_{a3}$, respectively, are given by

$$v_{a1} = \frac{1}{2\pi\varepsilon_0}\left(q_a \ln\frac{1}{r} + q_b \ln\frac{1}{D_{12}} + q_c \ln\frac{1}{D_{31}}\right) \tag{2.60}$$

$$v_{a2} = \frac{1}{2\pi\varepsilon_0}\left(q_a \ln\frac{1}{r} + q_b \ln\frac{1}{D_{23}} + q_c \ln\frac{1}{D_{12}}\right) \tag{2.61}$$

$$v_{a3} = \frac{1}{2\pi\varepsilon_0}\left(q_a \ln\frac{1}{r} + q_b \ln\frac{1}{D_{31}} + q_c \ln\frac{1}{D_{23}}\right) \tag{2.62}$$

The potential of conductor $a$ is assumed to be given by the following equation [3].

$$v_a = \frac{1}{3}(v_{a1} + v_{a2} + v_{a3}) \tag{2.63}$$

From (2.56) and (2.60) to (2.63), the capacitance in each phase is

$$C = \frac{2\pi\varepsilon_0 l}{\ln\left[(D_{12}D_{23}D_{31})^{1/3}/r\right]} \tag{2.64}$$

## 2.2.2.2 Composite Conductors

Consider the transmission line with composite conductors shown in Fig. 2.12. It is assumed that the three phases are transposed if not placed symmetrically. Let the composite conductors of phases $a$, $b$, $c$ occupy positions 1, 2, 3, respectively, in the first section, positions 2, 3, 1, respectively, in the second section, and positions 3, 1, 2, respectively, in the third section. Let each individual conductor of a composite conductor occupy each of the $n$ positions for equal lengths along a section. Let the charge per unit length on the individual conductors of phases $a$, $b$, and $c$ be $q_a/n$, $q_b/n$, and $q_c/n$, respectively, which satisfy (2.56). It is assumed that the charge per unit length on all individual conductors is same along the entire length of the transmission line. The potential of phase $a$ individual conductor at position $1k$ is

$$v_{a1k} = \frac{1}{2\pi\varepsilon_0 n} \sum_{m=1}^{n} \left[ q_a \ln \frac{1}{D'_{1k1m}} + q_b \ln \frac{1}{D'_{1k2m}} + q_c \ln \frac{1}{D'_{1k3m}} \right] \qquad (2.65)$$

where $D'_{pkqm}$ ($p$ and $q$ are 1, 2, or 3, and $pk \neq qm$) is the distance between the axes of conductors at positions $pk$ and $qm$; $D'_{pkpk}$ is the radius of the individual conductors. The potential of phase $a$ composite conductor in section 1, $v_{a1}$, is assumed to be equal to the average of the potentials of the individual conductors.

$$v_{a1} = \frac{1}{2\pi\varepsilon_0 n^2} \sum_{k=1}^{n} \sum_{m=1}^{n} \left[ q_a \ln \frac{1}{D'_{1k1m}} + q_b \ln \frac{1}{D'_{1k2m}} + q_c \ln \frac{1}{D'_{1k3m}} \right] \qquad (2.66)$$

The potential of phase $a$ composite conductor in sections 2 and 3, $v_{a2}$ and $v_{a3}$, respectively, are given by

$$v_{a2} = \frac{1}{2\pi\varepsilon_0 n^2} \sum_{k=1}^{n} \sum_{m=1}^{n} \left[ q_a \ln \frac{1}{D'_{2k2m}} + q_b \ln \frac{1}{D'_{2k3m}} + q_c \ln \frac{1}{D'_{2k1m}} \right] \qquad (2.67)$$

$$v_{a3} = \frac{1}{2\pi\varepsilon_0 n^2} \sum_{k=1}^{n} \sum_{m=1}^{n} \left[ q_a \ln \frac{1}{D'_{3k3m}} + q_b \ln \frac{1}{D'_{3k1m}} + q_c \ln \frac{1}{D'_{3k2m}} \right] \qquad (2.68)$$

The potential of phase $a$ composite conductor is assumed to be given by

$$v_a = \frac{1}{3}(v_{a1} + v_{a2} + v_{a3}) \qquad (2.69)$$

From (2.56) and (2.66) to (2.69), the capacitance in each phase is

$$C = \frac{2\pi\varepsilon_0 l}{\ln\left(D_m/D'_s\right)} \qquad (2.70)$$

**Fig. 2.13** Conductors and their images

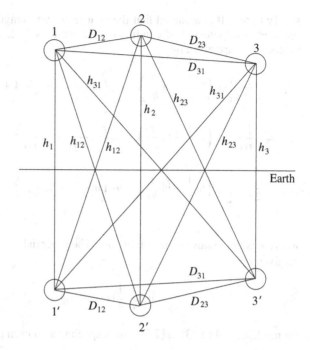

where

$$D_m \triangleq \left( \prod_{k=1}^{n} \prod_{m=1}^{n} D'_{1k2m} D'_{2k3m} D'_{3k1m} \right)^{1/(3n^2)} \tag{2.71}$$

$$D'_s \triangleq \left( \prod_{k=1}^{n} \prod_{m=1}^{n} D'_{1k1m} D'_{2k2m} D'_{3k3m} \right)^{1/(3n^2)} \tag{2.72}$$

### 2.2.2.3 Effect of Earth

The earth affects the distribution of the electric field due to charge on a conductor. The earth is at zero potential. The effect of earth is same as that of the image of the conductor [3]. The image of the conductor with charge $q_a$ per unit length is a conductor with charge $-q_a$ per unit length located at the same distance from the earth's surface below it as shown in Fig. 2.13. The images of conductors at positions 1, 2, and 3 are at positions $1'$, $2'$, and $3'$, respectively.

Transposition of conductors is assumed. Let the conductors $a, b, c$ occupy positions 1, 2, 3, respectively, in the first section, positions 2, 3, 1, respectively, in the second section, and positions 3, 1, 2, respectively, in the third section. Let the charge per unit length on the conductors $a, b,$ and $c$ be $q_a, q_b,$ and $q_c$, respectively, which satisfy (2.56). It is assumed that the charge per unit length is same in all the three sections. The potential of conductor $a$ in sections 1, 2, and 3, $v_{a1}$, $v_{a2}$, and $v_{a3}$, respectively, are given by

$$v_{a1} = \frac{1}{2\pi\varepsilon_0}\left[q_a\left(\ln\frac{1}{r} - \ln\frac{1}{h_1}\right) + q_b\left(\ln\frac{1}{D_{12}} - \ln\frac{1}{h_{12}}\right) + q_c\left(\ln\frac{1}{D_{31}} - \ln\frac{1}{h_{31}}\right)\right]$$

(2.73)

$$v_{a2} = \frac{1}{2\pi\varepsilon_0}\left[q_a\left(\ln\frac{1}{r} - \ln\frac{1}{h_2}\right) + q_b\left(\ln\frac{1}{D_{23}} - \ln\frac{1}{h_{23}}\right) + q_c\left(\ln\frac{1}{D_{12}} - \ln\frac{1}{h_{12}}\right)\right]$$

(2.74)

$$v_{a3} = \frac{1}{2\pi\varepsilon_0}\left[q_a\left(\ln\frac{1}{r} - \ln\frac{1}{h_3}\right) + q_b\left(\ln\frac{1}{D_{31}} - \ln\frac{1}{h_{31}}\right) + q_c\left(\ln\frac{1}{D_{23}} - \ln\frac{1}{h_{23}}\right)\right]$$

(2.75)

where $r$ is the radius of the conductors. The potential of conductor $a$ is assumed to be given by

$$v_a = \frac{1}{3}(v_{a1} + v_{a2} + v_{a3})$$

(2.76)

From (2.56) and (2.73) to (2.76), the capacitance in each phase is

$$C = \frac{2\pi\varepsilon_0 l}{\ln\dfrac{(D_{12}D_{23}D_{31})^{1/3}}{r} - \ln\left(\dfrac{h_{12}h_{23}h_{31}}{h_1h_2h_3}\right)^{1/3}}$$

(2.77)

Since $h_{12}h_{23}h_{31} > h_1h_2h_3$, the effect of earth is to increase the capacitance.

### 2.2.3  Transmission Line Model

Let $v_a$, $v_b$, and $v_c$ be the voltages with respect to the neutral, $i_a$, $i_b$, and $i_c$ be the currents, at the point which is at distance $x$ from the receiving end, as shown in Fig. 2.14. The currents satisfy (2.23). Then, the voltages and currents are related by the following equations.

$$\frac{\partial v_a}{\partial x} = \frac{R}{l}i_a + \frac{L}{l}\frac{\partial i_a}{\partial t}$$

(2.78)

$$\frac{\partial v_b}{\partial x} = \frac{R}{l}i_b + \frac{L}{l}\frac{\partial i_b}{\partial t}$$

(2.79)

$$\frac{\partial v_c}{\partial x} = \frac{R}{l}i_c + \frac{L}{l}\frac{\partial i_c}{\partial t}$$

(2.80)

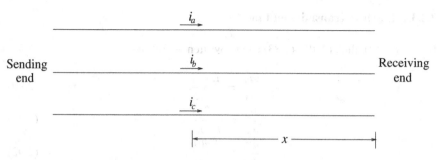

**Fig. 2.14** Transmission line

$$\frac{\partial i_a}{\partial x} = \frac{G}{l} v_a + \frac{C}{l} \frac{\partial v_a}{\partial t} \tag{2.81}$$

$$\frac{\partial i_b}{\partial x} = \frac{G}{l} v_b + \frac{C}{l} \frac{\partial v_b}{\partial t} \tag{2.82}$$

$$\frac{\partial i_c}{\partial x} = \frac{G}{l} v_c + \frac{C}{l} \frac{\partial v_c}{\partial t} \tag{2.83}$$

The equations for the three phases are decoupled.

Dividing (2.78)–(2.80) by $V_B$, and (2.81)–(2.83) by $I_B$ gives the equations in per unit quantities.

$$\frac{\partial \bar{v}_a}{\partial x} = \frac{\overline{R}}{l} \bar{i}_a + \frac{\overline{X}}{l\omega_B} \frac{\partial \bar{i}_a}{\partial t} \tag{2.84}$$

$$\frac{\partial \bar{v}_b}{\partial x} = \frac{\overline{R}}{l} \bar{i}_b + \frac{\overline{X}}{l\omega_B} \frac{\partial \bar{i}_b}{\partial t} \tag{2.85}$$

$$\frac{\partial \bar{v}_c}{\partial x} = \frac{\overline{R}}{l} \bar{i}_c + \frac{\overline{X}}{l\omega_B} \frac{\partial \bar{i}_c}{\partial t} \tag{2.86}$$

$$\frac{\partial \bar{i}_a}{\partial x} = \frac{\overline{G}}{l} \bar{v}_a + \frac{\overline{B}}{l\omega_B} \frac{\partial \bar{v}_a}{\partial t} \tag{2.87}$$

$$\frac{\partial \bar{i}_b}{\partial x} = \frac{\overline{G}}{l} \bar{v}_b + \frac{\overline{B}}{l\omega_B} \frac{\partial \bar{v}_b}{\partial t} \tag{2.88}$$

$$\frac{\partial \bar{i}_c}{\partial x} = \frac{\overline{G}}{l} \bar{v}_c + \frac{\overline{B}}{l\omega_B} \frac{\partial \bar{v}_c}{\partial t} \tag{2.89}$$

where $X \triangleq \omega_B L$, $B \triangleq \omega_B C$, base admittance $Y_B \triangleq 1/Z_B$, and base capacitance $C_B \triangleq Y_B/\omega_B$. Two special cases: lossless transmission line and sinusoidal operation, are considered.

### 2.2.3.1 Lossless Transmission Line

If $R = G = 0$, then (2.78)–(2.83) can be written as follows.

$$\frac{\partial v_a}{\partial x} = \frac{L}{l}\frac{\partial i_a}{\partial t} \qquad (2.90)$$

$$\frac{\partial i_a}{\partial x} = \frac{C}{l}\frac{\partial v_a}{\partial t} \qquad (2.91)$$

$$\frac{\partial v_b}{\partial x} = \frac{L}{l}\frac{\partial i_b}{\partial t} \qquad (2.92)$$

$$\frac{\partial i_b}{\partial x} = \frac{C}{l}\frac{\partial v_b}{\partial t} \qquad (2.93)$$

$$\frac{\partial v_c}{\partial x} = \frac{L}{l}\frac{\partial i_c}{\partial t} \qquad (2.94)$$

$$\frac{\partial i_c}{\partial x} = \frac{C}{l}\frac{\partial v_c}{\partial t} \qquad (2.95)$$

If the subscripts $a$, $b$, and $c$ are not shown, the solution for any phase is

$$i(x, t) = -f_1(x - v_p t) - f_2(x + v_p t) \qquad (2.96)$$
$$v(x, t) = Z_c f_1(x - v_p t) - Z_c f_2(x + v_p t) \qquad (2.97)$$

where $v_p \triangleq 1/\sqrt{LC}$ and $Z_c \triangleq \sqrt{L/C}$ [4]. $v_p$ is called phase velocity and $Z_c$ is called characteristic impedance. $f_1$ and $f_2$ are functions of $x$ and $t$. Let subscripts $S$ and $R$ denote sending end quantities and receiving end quantities, respectively. If only terminal response is of interest, the following method known as Bergeron's method is used. From (2.96) and (2.97),

$$i_R(t) = i(0, t) = -f_1(-v_p t) - f_2(v_p t) \qquad (2.98)$$
$$v_R(t) = v(0, t) = Z_c f_1(-v_p t) - Z_c f_2(v_p t) \qquad (2.99)$$
$$i_S\left(t - \frac{l}{v_p}\right) = i\left(l, t - \frac{l}{v_p}\right) = -f_1\left(2l - v_p t\right) - f_2\left(v_p t\right) \qquad (2.100)$$
$$v_S\left(t - \frac{l}{v_p}\right) = v\left(l, t - \frac{l}{v_p}\right) = Z_c f_1\left(2l - v_p t\right) - Z_c f_2\left(v_p t\right) \qquad (2.101)$$

Elimination of $f_1\left(-v_p t\right)$, $f_1\left(2l - v_p t\right)$, and $f_2(v_p t)$ from (2.98) to (2.101) gives

$$i_R(t) = i_S\left(t - \frac{l}{v_p}\right) + \frac{1}{Z_c}v_S\left(t - \frac{l}{v_p}\right) - \frac{1}{Z_c}v_R(t) \qquad (2.102)$$

This equation relates the receiving end current and voltage. Similarly, one can obtain the following equation which relates the sending end current and voltage.

$$i_S(t) = i_R \left( t - \frac{l}{v_p} \right) - \frac{1}{Z_c} v_R \left( t - \frac{l}{v_p} \right) + \frac{1}{Z_c} v_S(t) \qquad (2.103)$$

Dividing (2.102) and (2.103) by $I_B$ gives the equations in per unit quantities.

$$\bar{i}_R(t) = \bar{i}_S \left( t - \frac{l}{v_p} \right) + \frac{1}{\bar{Z}_c} \bar{v}_S \left( t - \frac{l}{v_p} \right) - \frac{1}{\bar{Z}_c} \bar{v}_R(t) \qquad (2.104)$$

$$\bar{i}_S(t) = \bar{i}_R \left( t - \frac{l}{v_p} \right) - \frac{1}{\bar{Z}_c} \bar{v}_R \left( t - \frac{l}{v_p} \right) + \frac{1}{\bar{Z}_c} \bar{v}_S(t) \qquad (2.105)$$

### 2.2.3.2 Sinusoidal Operation

Let the voltages and currents be sinusoidal with angular frequency $\omega_o$. Then voltages and currents can be represented by phasors. Let $\mathbf{V}$ and $\mathbf{I}$ be the notations for phasor representation of $v$ and $i$, respectively. If subscripts $a$, $b$, and $c$ are not shown, (2.78)–(2.83) can be written in the following form for each phase.

$$\frac{d\mathbf{V}}{dx} = \frac{R + j\omega_o L}{l} \mathbf{I} \qquad (2.106)$$

$$\frac{d\mathbf{I}}{dx} = \frac{G + j\omega_o C}{l} \mathbf{V} \qquad (2.107)$$

If $\mathbf{V}_R$ and $\mathbf{I}_R$ are the receiving end voltage and current, respectively, the solution of (2.106) and (2.107) is

$$\mathbf{V} = \cosh(\gamma x)\mathbf{V}_R + Z_c \sinh(\gamma x)\mathbf{I}_R \qquad (2.108)$$

$$\mathbf{I} = \frac{1}{Z_c} \sinh(\gamma x)\mathbf{V}_R + \cosh(\gamma x)\mathbf{I}_R \qquad (2.109)$$

where $Z_c$ is called characteristic impedance and $\gamma$ is called propagation constant.

$$Z_c \triangleq \sqrt{\frac{R + j\omega_o L}{G + j\omega_o C}}, \gamma \triangleq \frac{\sqrt{(R + j\omega_o L)(G + j\omega_o C)}}{l} \qquad (2.110)$$

If $\mathbf{V}_S$ and $\mathbf{I}_S$ are the sending end voltage and current, respectively, then from (2.108) and (2.109),

$$\mathbf{V}_S = \cosh(\gamma l)\mathbf{V}_R + Z_c \sinh(\gamma l)\mathbf{I}_R \qquad (2.111)$$

$$\mathbf{I}_S = \frac{1}{Z_c} \sinh(\gamma l)\mathbf{V}_R + \cosh(\gamma l)\mathbf{I}_R \qquad (2.112)$$

**Fig. 2.15** Equivalent $\pi$ circuit
of transmission line

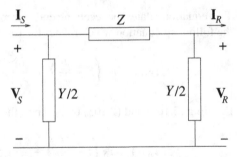

**Fig. 2.16** Nominal $\pi$ circuit
of transmission line

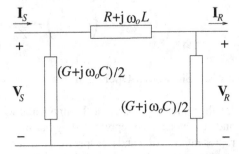

Each phase of the transmission line can be represented by the equivalent $\pi$ circuit shown in Fig. 2.15, where $Z$ is impedance and $Y$ is admittance. For this circuit, the following equations can be written.

$$\mathbf{V}_S = \left(1 + \frac{YZ}{2}\right)\mathbf{V}_R + Z\mathbf{I}_R \tag{2.113}$$

$$\mathbf{I}_S = Y\left(1 + \frac{YZ}{4}\right)\mathbf{V}_R + \left(1 + \frac{YZ}{2}\right)\mathbf{I}_R \tag{2.114}$$

Equating the coefficients of $\mathbf{V}_R$ and $\mathbf{I}_R$ in (2.111) and (2.113) gives

$$Z = (R + j\omega_o L)\frac{\sinh(\gamma l)}{\gamma l}, \; Y = (G + j\omega_o C)\frac{\tanh(\gamma l/2)}{\gamma l/2} \tag{2.115}$$

It is to be noted that

$$\lim_{l \to 0} Z = R + j\omega_o L, \; \lim_{l \to 0} Y = G + j\omega_o C \tag{2.116}$$

If $Z$ and $Y$ in Fig. 2.15 are replaced by the values of their respective limits as $l \to 0$, the circuit shown in Fig. 2.16 is obtained. This circuit is called nominal $\pi$ circuit. For transmission lines of length less than 240 km, the nominal $\pi$ circuit shown in Fig. 2.16 is a good approximation [3].

## 2.3 Kron's Transformation

Kron's transformation does a transformation of the three-phase voltages and currents as follows [5].

$$
\begin{bmatrix} v_D \\ v_Q \\ v_0 \end{bmatrix} \triangleq T_K \begin{bmatrix} v_a \\ v_b \\ v_c \end{bmatrix}, \quad \begin{bmatrix} i_D \\ i_Q \\ i_0 \end{bmatrix} \triangleq T_K \begin{bmatrix} i_a \\ i_b \\ i_c \end{bmatrix} \tag{2.117}
$$

where

$$
T_K \triangleq \frac{1}{\sqrt{3}} \begin{bmatrix} \sqrt{2}\cos(\omega_o t) & \sqrt{2}\cos(\omega_o t - 2\pi/3) & \sqrt{2}\cos(\omega_o t + 2\pi/3) \\ \sqrt{2}\sin(\omega_o t) & \sqrt{2}\sin(\omega_o t - 2\pi/3) & \sqrt{2}\sin(\omega_o t + 2\pi/3) \\ 1 & 1 & 1 \end{bmatrix} \tag{2.118}
$$

where $\omega_o$ is the operating frequency. It can be verified that $T_K^{-1} = T_K^T$.

If $v_0 = i_0 = 0$, then (2.117) can be written as

$$
\begin{bmatrix} v_D \\ v_Q \end{bmatrix} = T_K' \begin{bmatrix} v_a \\ v_b \\ v_c \end{bmatrix}, \quad \begin{bmatrix} i_D \\ i_Q \end{bmatrix} = T_K' \begin{bmatrix} i_a \\ i_b \\ i_c \end{bmatrix} \tag{2.119}
$$

where

$$
T_K' \triangleq \frac{1}{\sqrt{3}} \begin{bmatrix} \sqrt{2}\cos(\omega_o t) & \sqrt{2}\cos(\omega_o t - 2\pi/3) & \sqrt{2}\cos(\omega_o t + 2\pi/3) \\ \sqrt{2}\sin(\omega_o t) & \sqrt{2}\sin(\omega_o t - 2\pi/3) & \sqrt{2}\sin(\omega_o t + 2\pi/3) \end{bmatrix} \tag{2.120}
$$

In certain studies, high-frequency transients in the transformer and the transmission line are neglected. Then, Kron's transformation results in simplification of equations. Kron's transformation also enables generalization of the definitions of certain electrical quantities.

### 2.3.1 Definitions

There are quantities such as voltage magnitude, phase angle, frequency, reactive power etc. which are well defined in steady state when voltages and currents are sinusoidally varying and balanced. The definition of these quantities will be generalized so that they can be used even in the presence of harmonics and during a transient when the voltage and current are not sinusoidal; however, these definitions are made with the assumption that $v_0 = i_0 = 0$.

Consider the shunt-connected equipment shown in Fig. 2.17. Let $v_a$, $v_b$, and $v_c$ be the voltages of terminals $a$, $b$, and $c$, respectively, with respect to the neutral.

**Fig. 2.17** Shunt-connected
equipment

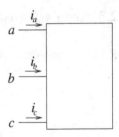

The magnitude $V$ and phase angle $\phi$ of the voltage of the three-phase bus, at which the equipment in Fig. 2.17 is connected, are defined as

$$V \triangleq \sqrt{v_D^2 + v_Q^2} \tag{2.121}$$

$$\phi \triangleq \tan^{-1} \frac{v_D}{v_Q} \tag{2.122}$$

In other words, $V \angle \phi = v_Q + j v_D$. $v_Q$ and $v_D$ can be expressed in terms of $V$ and $\phi$ as follows.

$$v_Q = V \cos \phi, \ v_D = V \sin \phi \tag{2.123}$$

If $v_a$, $v_b$, and $v_c$ are obtained from these expressions for $v_Q$ and $v_D$ using (2.119), then

$$v_a = \sqrt{\frac{2}{3}} V \sin(\omega_o t + \phi) \tag{2.124}$$

$$v_b = \sqrt{\frac{2}{3}} V \sin\left(\omega_o t + \phi - \frac{2\pi}{3}\right) \tag{2.125}$$

$$v_c = \sqrt{\frac{2}{3}} V \sin\left(\omega_o t + \phi + \frac{2\pi}{3}\right) \tag{2.126}$$

Therefore, if $v_a$, $v_b$, and $v_c$ are sinusoidal with angular frequency $\omega_o$ and balanced, $V$ is the rms value of the line-to-line voltage and $\phi$ is the phase angle of $v_a$.

The frequency at the three-phase bus $f$ is defined as

$$f \triangleq f_o + \frac{1}{2\pi} \frac{d\phi}{dt} \tag{2.127}$$

where $f_o \triangleq \omega_o/(2\pi)$.

Similar to voltage magnitude and phase angle definitions, the magnitude $I$ and phase angle $\psi$ of the current drawn by the equipment in Fig. 2.17 are defined as

$$I \triangleq \sqrt{i_D^2 + i_Q^2} \tag{2.128}$$

$$\psi \triangleq \tan^{-1} \frac{i_D}{i_Q} \tag{2.129}$$

If $i_a$, $i_b$, and $i_c$ are sinusoidal with angular frequency $\omega_o$ and balanced, then $I$ is $\sqrt{3}$ times the rms value of $i_a$, $i_b$, or $i_c$, and $\psi$ is the phase angle of $i_a$.

The power drawn by the equipment in Fig. 2.17 is

$$P = v_a i_a + v_b i_b + v_c i_c \tag{2.130}$$

From (2.119),

$$P = v_D i_D + v_Q i_Q \tag{2.131}$$

It can be seen that

$$P = \text{Re}[V \angle \phi I \angle (-\psi)] \tag{2.132}$$

$P$ is also known as active power. The reactive power $Q$ drawn by the equipment in Fig. 2.17 is defined as

$$Q \triangleq \text{Im}[V \angle \phi I \angle (-\psi)] = v_D i_Q - v_Q i_D \tag{2.133}$$

The active current $i_A$ and the reactive current $i_R$ drawn by the equipment in Fig. 2.17 are defined as

$$i_A \triangleq I \cos(\phi - \psi) = i_Q \cos \phi + i_D \sin \phi \tag{2.134}$$

$$i_R \triangleq I \sin(\phi - \psi) = i_Q \sin \phi - i_D \cos \phi \tag{2.135}$$

It is to be noted that $i_A > 0 \Leftrightarrow P > 0$, and $i_R > 0 \Leftrightarrow Q > 0$. The reactive current is said to be inductive if it is positive, and is said to be capacitive if it is negative.

Consider the series-connected equipment shown in Fig. 2.18. The magnitude $V$ and phase angle $\phi$ of the voltage across the equipment in Fig. 2.18 are given by (2.119)–(2.122) using $v_a$, $v_b$, and $v_c$ of Fig. 2.18. Similarly, the magnitude $I$ and phase angle $\psi$ of the current through the equipment in Fig. 2.18 are given by (2.119), (2.120), (2.128), and (2.129) using $i_a$, $i_b$, and $i_c$ of Fig. 2.18. The active voltage $v_A$ and the reactive voltage $v_R$ across the equipment in Fig. 2.18 are defined as

$$v_A \triangleq V \cos(\phi - \psi) \tag{2.136}$$

$$v_R \triangleq V \sin(\phi - \psi) \tag{2.137}$$

**Fig. 2.18** Series-connected
equipment

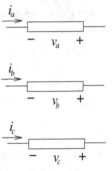

If $v_A$ is positive, active power is supplied by the equipment, otherwise, active power
is drawn by the equipment. The reactive voltage is said to be capacitive if it is positive
and inductive if it is negative.

### 2.3.2 Application to Transformer

Equations (2.18)–(2.20) of the wye-wye-connected transformer can be written as

$$
\begin{bmatrix} \bar{v}_{2a} \\ \bar{v}_{2b} \\ \bar{v}_{2c} \end{bmatrix} = \begin{bmatrix} \bar{v}_{1a} \\ \bar{v}_{1b} \\ \bar{v}_{1c} \end{bmatrix} - (\bar{R}_1 + \bar{R}_2) \begin{bmatrix} \bar{i}_a \\ \bar{i}_b \\ \bar{i}_c \end{bmatrix} - \frac{1}{\omega_B}(\bar{X}_1 + \bar{X}_2) \begin{bmatrix} d\bar{i}_a/dt \\ d\bar{i}_b/dt \\ d\bar{i}_c/dt \end{bmatrix} \quad (2.138)
$$

By Kron's transformation,

$$
T_K'^T \begin{bmatrix} \bar{v}_{2D} \\ \bar{v}_{2Q} \end{bmatrix} = T_K'^T \begin{bmatrix} \bar{v}_{1D} \\ \bar{v}_{1Q} \end{bmatrix} - (\bar{R}_1 + \bar{R}_2) T_K'^T \begin{bmatrix} \bar{i}_D \\ \bar{i}_Q \end{bmatrix} - \frac{1}{\omega_B}(\bar{X}_1 + \bar{X}_2) \frac{d}{dt}\left( T_K'^T \begin{bmatrix} \bar{i}_D \\ \bar{i}_Q \end{bmatrix} \right)
$$
$$
(2.139)
$$

where $\begin{bmatrix} \bar{v}_{1D} \\ \bar{v}_{1Q} \end{bmatrix} \triangleq T_K' \begin{bmatrix} \bar{v}_{1a} \\ \bar{v}_{1b} \\ \bar{v}_{1c} \end{bmatrix}$, $\begin{bmatrix} \bar{v}_{2D} \\ \bar{v}_{2Q} \end{bmatrix} \triangleq T_K' \begin{bmatrix} \bar{v}_{2a} \\ \bar{v}_{2b} \\ \bar{v}_{2c} \end{bmatrix}$ and $\begin{bmatrix} \bar{i}_D \\ \bar{i}_Q \end{bmatrix} \triangleq T_K' \begin{bmatrix} \bar{i}_a \\ \bar{i}_b \\ \bar{i}_c \end{bmatrix}$.

Pre-multiplying (2.139) by $T_K'$ gives

$$
\bar{v}_{2D} = \bar{v}_{1D} - (\bar{R}_1 + \bar{R}_2)\bar{i}_D - \frac{\omega_o}{\omega_B}(\bar{X}_1 + \bar{X}_2)\bar{i}_Q - \frac{1}{\omega_B}(\bar{X}_1 + \bar{X}_2)\frac{d\bar{i}_D}{dt} \quad (2.140)
$$

**Fig. 2.19** Equivalent $\pi$ circuit
of phase $a$ of transmission line

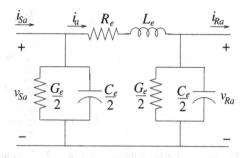

$$\bar{v}_{2Q} = \bar{v}_{1Q} - (\overline{R}_1 + \overline{R}_2)\bar{i}_Q + \frac{\omega_o}{\omega_B}\left(\overline{X}_1 + \overline{X}_2\right)\bar{i}_D - \frac{1}{\omega_B}\left(\overline{X}_1 + \overline{X}_2\right)\frac{d\bar{i}_Q}{dt} \quad (2.141)$$

If the high-frequency transients are to be neglected, then the last term on the right-hand side of (2.140) and (2.141) are set to zero. For balanced sinusoidal operation at angular frequency $\omega_o$, all transformed variables are constant and hence the last term on the right-hand side of (2.140) and (2.141) is equal to zero. The factor $\omega_o/\omega_B$ in one of the terms of (2.140) and (2.141) is usually approximated to 1. Therefore,

$$\bar{v}_{2D} = \bar{v}_{1D} - (\overline{R}_1 + \overline{R}_2)\bar{i}_D - \left(\overline{X}_1 + \overline{X}_2\right)\bar{i}_Q \quad (2.142)$$

$$\bar{v}_{2Q} = \bar{v}_{1Q} - (\overline{R}_1 + \overline{R}_2)\bar{i}_Q + \left(\overline{X}_1 + \overline{X}_2\right)\bar{i}_D \quad (2.143)$$

### 2.3.3 Application to Transmission Line

For sinusoidal operation, the equivalent $\pi$ circuit of the transmission line shown in Fig. 2.15 is applicable. Let $Z = R_e + j\omega_o L_e$ and $Y = G_e + j\omega_o C_e$. The circuit of Fig. 2.15 can be redrawn as shown in Fig. 2.19 for phase $a$.

From the circuit diagram in Fig. 2.19,

$$v_{Sa} - v_{Ra} = R_e i_a + L_e \frac{di_a}{dt} \quad (2.144)$$

$$i_{Sa} - i_a = \frac{G_e}{2} v_{Sa} + \frac{C_e}{2} \frac{dv_{Sa}}{dt} \quad (2.145)$$

$$i_a - i_{Ra} = \frac{G_e}{2} v_{Ra} + \frac{C_e}{2} \frac{dv_{Ra}}{dt} \quad (2.146)$$

Similarly, for phases $b$ and $c$,

$$v_{Sb} - v_{Rb} = R_e i_b + L_e \frac{di_b}{dt} \quad (2.147)$$

$$v_{Sc} - v_{Rc} = R_e i_c + L_e \frac{di_c}{dt} \quad (2.148)$$

$$i_{Sb} - i_b = \frac{G_e}{2}v_{Sb} + \frac{C_e}{2}\frac{dv_{Sb}}{dt} \tag{2.149}$$

$$i_{Sc} - i_c = \frac{G_e}{2}v_{Sc} + \frac{C_e}{2}\frac{dv_{Sc}}{dt} \tag{2.150}$$

$$i_b - i_{Rb} = \frac{G_e}{2}v_{Rb} + \frac{C_e}{2}\frac{dv_{Rb}}{dt} \tag{2.151}$$

$$i_c - i_{Rc} = \frac{G_e}{2}v_{Rc} + \frac{C_e}{2}\frac{dv_{Rc}}{dt} \tag{2.152}$$

By Kron's transformation, (2.144)–(2.152) can be written as

$$v_{SD} - v_{RD} = R_e i_D + \omega_o L_e i_Q + L_e\frac{di_D}{dt} \tag{2.153}$$

$$v_{SQ} - v_{RQ} = R_e i_Q - \omega_o L_e i_D + L_e\frac{di_Q}{dt} \tag{2.154}$$

$$i_{SD} - i_D = \frac{G_e}{2}v_{SD} + \omega_o\frac{C_e}{2}v_{SQ} + \frac{C_e}{2}\frac{dv_{SD}}{dt} \tag{2.155}$$

$$i_{SQ} - i_Q = \frac{G_e}{2}v_{SQ} - \omega_o\frac{C_e}{2}v_{SD} + \frac{C_e}{2}\frac{dv_{SQ}}{dt} \tag{2.156}$$

$$i_D - i_{RD} = \frac{G_e}{2}v_{RD} + \omega_o\frac{C_e}{2}v_{RQ} + \frac{C_e}{2}\frac{dv_{RD}}{dt} \tag{2.157}$$

$$i_Q - i_{RQ} = \frac{G_e}{2}v_{RQ} - \omega_o\frac{C_e}{2}v_{RD} + \frac{C_e}{2}\frac{dv_{RQ}}{dt} \tag{2.158}$$

where $\begin{bmatrix} v_{SD} \\ v_{SQ} \end{bmatrix} \triangleq T'_K \begin{bmatrix} v_{Sa} \\ v_{Sb} \\ v_{Sc} \end{bmatrix}$, $\begin{bmatrix} v_{RD} \\ v_{RQ} \end{bmatrix} \triangleq T'_K \begin{bmatrix} v_{Ra} \\ v_{Rb} \\ v_{Rc} \end{bmatrix}$, $\begin{bmatrix} i_{SD} \\ i_{SQ} \end{bmatrix} \triangleq T'_K \begin{bmatrix} i_{Sa} \\ i_{Sb} \\ i_{Sc} \end{bmatrix}$,

$\begin{bmatrix} i_{RD} \\ i_{RQ} \end{bmatrix} \triangleq T'_K \begin{bmatrix} i_{Ra} \\ i_{Rb} \\ i_{Rc} \end{bmatrix}$, and $\begin{bmatrix} i_D \\ i_Q \end{bmatrix} \triangleq T'_K \begin{bmatrix} i_a \\ i_b \\ i_c \end{bmatrix}$. For balanced sinusoidal operation

at frequency $\omega_o$, all transformed variables are constant. Therefore, (2.153)–(2.158) can be written as

$$v_{SD} - v_{RD} = R_e i_D + \omega_o L_e i_Q \tag{2.159}$$

$$v_{SQ} - v_{RQ} = R_e i_Q - \omega_o L_e i_D \tag{2.160}$$

$$i_{SD} - i_D = \frac{G_e}{2}v_{SD} + \omega_o\frac{C_e}{2}v_{SQ} \tag{2.161}$$

$$i_{SQ} - i_Q = \frac{G_e}{2}v_{SQ} - \omega_o\frac{C_e}{2}v_{SD} \tag{2.162}$$

$$i_D - i_{RD} = \frac{G_e}{2}v_{RD} + \omega_o\frac{C_e}{2}v_{RQ} \tag{2.163}$$

$$i_Q - i_{RQ} = \frac{G_e}{2}v_{RQ} - \omega_o\frac{C_e}{2}v_{RD} \tag{2.164}$$

Dividing (2.159) and (2.160) by $V_B$, and (2.161)–(2.164) by $I_B$ gives the following equations in per unit quantities.

$$\bar{v}_{SD} - \bar{v}_{RD} = \bar{R}_e \bar{i}_D + \frac{\omega_o}{\omega_B} \bar{X}_e \bar{i}_Q \tag{2.165}$$

$$\bar{v}_{SQ} - \bar{v}_{RQ} = \bar{R}_e \bar{i}_Q - \frac{\omega_o}{\omega_B} \bar{X}_e \bar{i}_D \tag{2.166}$$

$$\bar{i}_{SD} - \bar{i}_D = \frac{\bar{G}_e}{2} \bar{v}_{SD} + \frac{\omega_o}{\omega_B} \frac{\bar{B}_e}{2} \bar{v}_{SQ} \tag{2.167}$$

$$\bar{i}_{SQ} - \bar{i}_Q = \frac{\bar{G}_e}{2} \bar{v}_{SQ} - \frac{\omega_o}{\omega_B} \frac{\bar{B}_e}{2} \bar{v}_{SD} \tag{2.168}$$

$$\bar{i}_D - \bar{i}_{RD} = \frac{\bar{G}_e}{2} \bar{v}_{RD} + \frac{\omega_o}{\omega_B} \frac{\bar{B}_e}{2} \bar{v}_{RQ} \tag{2.169}$$

$$\bar{i}_Q - \bar{i}_{RQ} = \frac{\bar{G}_e}{2} \bar{v}_{RQ} - \frac{\omega_o}{\omega_B} \frac{\bar{B}_e}{2} \bar{v}_{RD} \tag{2.170}$$

where $X_e \triangleq \omega_B L_e$ and $B_e \triangleq \omega_B C_e$. The factor $\omega_o/\omega_B$ in one of the terms in all equations is usually approximated to 1. As an approximation, (2.165)–(2.170) are used even during transients.

## 2.4 Load

In many system studies, the effects of the subtransmission and the distribution networks along with the connected load devices are represented by an aggregated load at a transmission substation. The load model is given by the expressions for active power $P$ and reactive power $Q$ drawn, in terms of voltage magnitude and/or frequency [5, 6]. Two commonly used models are:

- 
$$P = P_o \left(\frac{V}{V_o}\right)^a \left[1 + k_{pf}(f - f_o)\right] \tag{2.171}$$

$$Q = Q_o \left(\frac{V}{V_o}\right)^b \left[1 + k_{qf}(f - f_o)\right] \tag{2.172}$$

- 
$$P = P_o \left[p_1 \left(\frac{V}{V_o}\right)^2 + p_2 \frac{V}{V_o} + p_3\right] \left[1 + k_{pf}(f - f_o)\right] \tag{2.173}$$

$$Q = Q_o \left[q_1 \left(\frac{V}{V_o}\right)^2 + q_2 \frac{V}{V_o} + q_3\right] \left[1 + k_{qf}(f - f_o)\right] \tag{2.174}$$

Subscript $o$ identifies the values of the respective variables at the operating point. $a$, $b$, $p_1$, $p_2$, $p_3$, $q_1$, $q_2$, $q_3$, $k_{pf}$, and $k_{qf}$ are constants; $p_1 + p_2 + p_3 = 1$ and $q_1 + q_2 + q_3 = 1$. If frequency dependence is not to be considered, $k_{pf}$ and $k_{qf}$ are set to zero.

Equations (2.171)–(2.174) in per unit quantities are

$$\overline{P} = \frac{P_o V_B^a}{S_B V_o^a} \overline{V}^a \left[1 + k_{pf}(f - f_o)\right] \tag{2.175}$$

$$\overline{Q} = \frac{Q_o V_B^b}{S_B V_o^b} \overline{V}^b \left[1 + k_{qf}(f - f_o)\right] \tag{2.176}$$

$$\overline{P} = \frac{P_o}{S_B} \left[p_1 \frac{V_B^2}{V_o^2} \overline{V}^2 + p_2 \frac{V_B}{V_o} \overline{V} + p_3\right] \left[1 + k_{pf}(f - f_o)\right] \tag{2.177}$$

$$\overline{Q} = \frac{Q_o}{S_B} \left[q_1 \frac{V_B^2}{V_o^2} \overline{V}^2 + q_2 \frac{V_B}{V_o} \overline{V} + q_3\right] \left[1 + k_{qf}(f - f_o)\right] \tag{2.178}$$

# References

1. C.A. Gross, *Power System Analysis*, 2nd edn. (Wiley, New York, 1986)
2. S. Krishna, Teaching calculation of inductance of power transmission lines. Int. J. Electr. Eng. Educ. **48**(4), 434–443 (2011)
3. J.J. Grainger, W.D. Stevenson Jr, *Power System Analysis* (Tata McGraw-Hill, Noida, 1994)
4. P.W. Sauer, M.A. Pai, *Power System Dynamics and Stability* (Pearson Education, Upper Saddle River, 1998)
5. K.R. Padiyar, *Power System Dynamics: Stability and Control*, 2nd edn. (BS Publications, Hyderabad, 2002)
6. P. Kundur, *Power System Stability and Control* (Tata McGraw-Hill, Noida, 1994)

# Chapter 3
# DC and Flexible AC Transmission Systems

Under certain circumstances, it is advantageous to transmit power over a DC transmission line instead of an AC transmission line. DC transmission requires conversion from AC to DC at one end of the transmission line, and conversion from DC to AC at the other end of the transmission line. These conversions are done by a circuit called converter which consists of power semiconductor devices.

Flexible AC transmission system (FACTS) is an AC transmission system incorporating equipment, made up of power semiconductor devices, in order to enhance controllability and increase power transfer capability [1]. Such equipment is called a FACTS controller.

## 3.1 Power Semiconductor Devices

The power semiconductor devices are used as switches. A mechanical switch is said to be in the on state if it is conducting current and is said to be in the off state if it is not conducting current. Similarly, power semiconductor devices can be in the on state or the off state. The transition from off state to on state is called turn on, and the transition from on state to off state is called turn off. It will be assumed that these devices are ideal. In other words, the following assumptions are made:

1. The voltage across a device is zero when it is on.
2. The current through a device is zero when it is off.
3. The time taken to turn on and turn off is zero.

The power semiconductor devices can be classified into diode, thyristor, and controllable switches [2].

The diode has two terminals—anode (A) and cathode (K)—and its circuit symbol is shown in Fig. 3.1a. The diode conducts current from anode to cathode if the potential of anode is higher than that of cathode, and it does not conduct current if the anode potential is less than cathode potential. The state of the diode depends on the circuit conditions. Hence, the diode is said to be an uncontrolled device.

S Krishna, *An Introduction to Modelling of Power System Components*, 73
SpringerBriefs in Electrical and Computer Engineering,
DOI: 10.1007/978-81-322-1847-0_3, © The Author(s) 2014

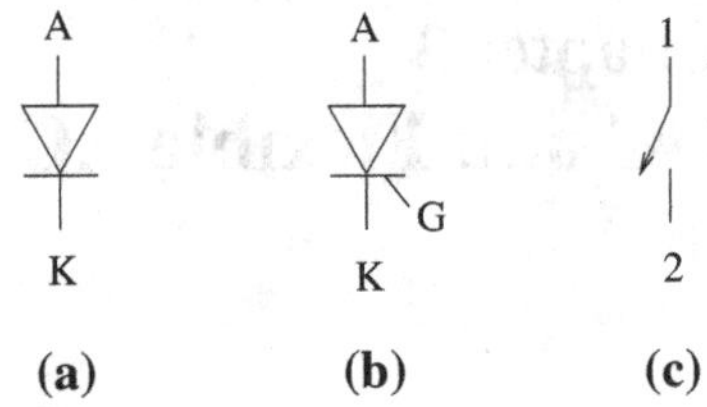

**Fig. 3.1** Power
semiconductor devices:
**a** diode, **b** thyristor, and
**c** controllable switch

(a)          (b)          (c)

The thyristor has three terminals—anode (A), cathode (K), and gate (G)—and its circuit symbol is shown in Fig. 3.1b. The thyristor starts conducting current from anode to cathode if the anode potential is higher than the cathode potential, and a pulse of positive current flows from gate to cathode. The thyristor stops conducting if the anode to cathode current goes to zero, and the cathode potential is higher than anode potential for a certain minimum duration. The gate is said to be the control terminal. Since the thyristor can be turned on by a control signal but cannot be turned off by a control signal, it is said to be a semi-controllable device.

The circuit symbol of a controllable switch is shown in Fig. 3.1c. The circuit symbol has two terminals—1 and 2—and the control terminal is not shown. The switch can be turned on by a control signal if the potential of terminal 1 is higher than that of terminal 2, and the current flows from terminal 1 to terminal 2. The switch can also be turned off by a control signal. Gate turn off (GTO) thyristor and insulated gate bipolar transistor (IGBT) are examples of controllable switch.

A device is said to be forward-biased if the potential of anode (or terminal 1) is higher than that of cathode (or terminal 2). On the other hand, the device is said to be reverse-biased if the potential of cathode (or terminal 2) is higher than that of anode (or terminal 1).

When used at the transmission level, many devices are connected in series and/or parallel to obtain the required voltage and/or current rating [1, 3]. This combination of devices is called a valve. It is assumed that a valve behaves as a single device; all the devices in a valve turn on and turn off simultaneously. The circuit symbol of a valve is the same as that of the device.

## 3.2 DC Transmission System

A DC transmission line requires two converters one at each end where one converter operates as a rectifier and the other operates as an inverter. A converter is said to operate as a rectifier if the average power flow is from the AC side to the DC side. If the average power flow is from the DC side to the AC side, the converter is said to operate as an inverter. The type of converter used for long-distance bulk power transmission is line-commutated converter [4].

**Fig. 3.2** Line-commutated converter

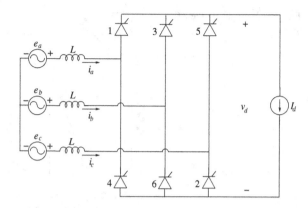

## 3.2.1 Line-Commutated Converter

The circuit diagram of a six-pulse line-commutated converter is shown in Fig. 3.2. The converter has six thyristor valves. The circuit on the AC side is represented by a three-phase balanced voltage source with an inductance $L$ in each phase. A large smoothing reactor is used on the DC side [4, 5]. Therefore, the circuit on the DC side is represented by a constant current source.

### 3.2.1.1 Two-Valve Conduction Mode

For the sake of starting with a simplified analysis, let $L = 0$. If the gate currents are continuously applied, the valves behave as diodes. At any instant, one valve among 1, 3, and 5 conducts and one valve among 2, 4, and 6 conducts. Among the valves 1, 3, and 5, the valve whose anode is at the highest potential conducts and the other two valves are reverse-biased. Among the valves 2, 4, and 6, the valve whose cathode is at the lowest potential conducts and the other two valves are reverse-biased. Let $e_a$, $e_b$, and $e_c$ be given by

$$e_a = \sqrt{\frac{2}{3}}V \sin(\omega_o t + 150°) \tag{3.1}$$

$$e_b = \sqrt{\frac{2}{3}}V \sin(\omega_o t + 30°) \tag{3.2}$$

$$e_c = \sqrt{\frac{2}{3}}V \sin(\omega_o t - 90°) \tag{3.3}$$

The plots of $e_a$, $e_b$, and $e_c$ are shown in Fig. 3.3. Each valve conducts for 120°. The valves turn on in the sequence 1, 2, 3, 4, 5, and 6. The duration between any two consecutive instants of turn on is 60°. One cycle of the AC voltage can be divided into six equal intervals where each interval corresponds to the conduction of a pair

**Fig. 3.3** Voltages on the AC side

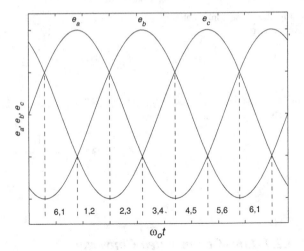

**Table 3.1** DC-side voltage and voltage across a valve for two-valve conduction mode

| Interval | Valves that conduct | $v_d$ | Voltage across valve 1 |
|---|---|---|---|
| $\alpha < \omega_o t < \alpha + 60°$ | 2, 3 | $e_b - e_c$ | $e_a - e_b$ |
| $\alpha + 60° < \omega_o t < \alpha + 120°$ | 3, 4 | $e_b - e_a$ | $e_a - e_b$ |
| $\alpha + 120° < \omega_o t < \alpha + 180°$ | 4, 5 | $e_c - e_a$ | $e_a - e_c$ |
| $\alpha + 180° < \omega_o t < \alpha + 240°$ | 5, 6 | $e_c - e_b$ | $e_a - e_c$ |
| $\alpha + 240° < \omega_o t < \alpha + 300°$ | 6, 1 | $e_a - e_b$ | 0 |
| $\alpha + 300° < \omega_o t < \alpha + 360°$ | 1, 2 | $e_a - e_c$ | 0 |

of valves. The six intervals and the conducting valves during each interval are shown in Fig. 3.3.

The instant of natural conduction of a valve is the instant at which the valve starts conducting if the gate current is continuously applied. For example, the instant of natural conduction of valve 3 is when $\omega_o t$ is equal to zero. Instead of continuous gate currents, gate current pulses are applied which are delayed by an angle $\alpha$ with respect to the instants of natural conduction; $\alpha$ is called delay angle. For example, valve 3 is turned on at $\omega_o t = \alpha$. Valve 1 stops conducting when valve 3 is turned on. The transfer of current from one valve to another is known as commutation. Voltage across valve 1 when valve 3 is turned on is $-\sqrt{2}V \sin(\omega_o t)$, and this voltage appears across valve 1 for $\alpha < \omega_o t < \alpha + 120°$. Immediately after valve 1 stops conducting, it takes some time for it to withstand positive voltage [2, 4, 6] and this time is given in terms of an angle denoted by $\xi_o$. Therefore, $\alpha$ can take a value between 0 and $180° - \xi_o$.

For each interval, the valves that conduct, DC voltage $v_d$, and voltage across valve 1 are given in Table 3.1.

The average DC voltage is

$$V_d = \frac{3\sqrt{2}}{\pi} V \cos \alpha \tag{3.4}$$

**Fig. 3.4** AC-side current

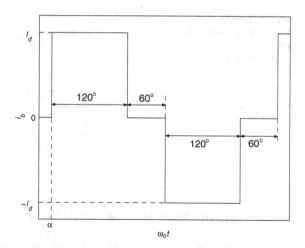

The converter operates as a rectifier if $0 \leq \alpha < 90°$ and as an inverter if $90° < \alpha < 180° - \xi_o$.

The harmonic components in the DC voltage are of order $h = 6k, k = 1, 2, 3, \ldots$ The rms value of the $h$th-order harmonic component is

$$V_h = \frac{6V}{\pi(h^2 - 1)} \left[1 + (h^2 - 1) \sin^2 \alpha\right]^{1/2} \tag{3.5}$$

The plot of an AC-side current is shown in Fig. 3.4.

The rms value of the fundamental component of the AC-side currents is

$$I_1 = \frac{\sqrt{6}}{\pi} I_d \tag{3.6}$$

The harmonic components in the AC-side currents are of order $h = 6k \pm 1, k = 1, 2, 3, \ldots$ The rms value of the $h$th-order harmonic component is

$$I_h = \frac{I_1}{h} \tag{3.7}$$

An angle known as angle of advance, denoted by $\beta$, is defined [4] as

$$\beta \triangleq 180° - \alpha \tag{3.8}$$

If $\alpha > 60°$, $\beta$ gives the duration for which the voltage across a valve is negative after it stops conducting; $\beta$ should be greater than $\xi_o$. An angle $\xi$ known as commutation margin angle is defined as the duration for which the voltage across a valve is negative after it stops conducting [4]. This angle is relevant for inverter operation for which

$$\xi = \beta \tag{3.9}$$

**Fig. 3.5** Effective circuit
in the first subinterval for
two- and three-valve
conduction mode

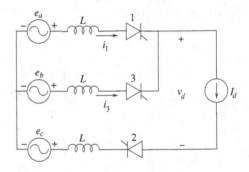

### 3.2.1.2 Two- and Three-Valve Conduction Mode

Due to the inductance $L$ on the AC side, the current in a valve can vary only at
a finite rate, and hence, commutation from one valve to another takes some time.
For example, when valve 3 is turned on, the current transfer from valve 1 to valve
3 requires a certain time period during which both valves conduct. This duration is
measured in terms of an angle known as overlap angle or commutation angle denoted
by $u$. For normal operation, $0 < u < 60°$.

One cycle of the AC voltage can be divided into six equal intervals. Each interval
can be divided into two subintervals. Three valves conduct in the first subinterval, and
two valves conduct in the second subinterval. For steady-state analysis, it is sufficient
to consider only one interval. The interval $\alpha < \omega_o t < \alpha + 60°$ is considered.

The effective circuit to be analyzed in the first subinterval ($\alpha < \omega_o t < \alpha + u$) is
shown in Fig. 3.5.

From the circuit diagram in Fig. 3.5,

$$L\frac{di_3}{dt} - L\frac{d(I_d - i_3)}{dt} = e_b - e_a \qquad (3.10)$$

Since $i_3(\alpha) = 0$, the solution of this equation is

$$i_3 = \frac{V}{\sqrt{2}\omega_o L}[\cos\alpha - \cos(\omega_o t)] \qquad (3.11)$$

Substituting $i_3(\alpha + u) = I_d$ gives

$$I_d = \frac{V}{\sqrt{2}\omega_o L}[\cos\alpha - \cos(\alpha + u)] \qquad (3.12)$$

From the circuit diagram in Fig. 3.5, using the relation $e_a + e_b + e_c = 0$,

$$v_d = -\frac{3}{2}e_c \qquad (3.13)$$

**Table 3.2** DC-side voltage and voltage across a valve for two- and three-valve conduction mode

| Subinterval | Valves that conduct | $v_d$ | Voltage across valve 1 |
|---|---|---|---|
| $\alpha < \omega_o t < \alpha + u$ | 1, 2, 3 | $-3e_c/2$ | 0 |
| $\alpha + u < \omega_o t < \alpha + 60°$ | 2, 3 | $e_b - e_c$ | $e_a - e_b$ |
| $\alpha + 60° < \omega_o t < \alpha + u + 60°$ | 2, 3, 4 | $3e_b/2$ | $-3e_b/2$ |
| $\alpha + u + 60° < \omega_o t < \alpha + 120°$ | 3, 4 | $e_b - e_a$ | $e_a - e_b$ |
| $\alpha + 120° < \omega_o t < \alpha + u + 120°$ | 3, 4, 5 | $-3e_a/2$ | $3e_a/2$ |
| $\alpha + u + 120° < \omega_o t < \alpha + 180°$ | 4, 5 | $e_c - e_a$ | $e_a - e_c$ |
| $\alpha + 180° < \omega_o t < \alpha + u + 180°$ | 4, 5, 6 | $3e_c/2$ | $-3e_c/2$ |
| $\alpha + u + 180° < \omega_o t < \alpha + 240°$ | 5, 6 | $e_c - e_b$ | $e_a - e_c$ |
| $\alpha + 240° < \omega_o t < \alpha + u + 240°$ | 5, 6, 1 | $-3e_b/2$ | 0 |
| $\alpha + u + 240° < \omega_o t < \alpha + 300°$ | 6, 1 | $e_a - e_b$ | 0 |
| $\alpha + 300° < \omega_o t < \alpha + u + 300°$ | 6, 1, 2 | $3e_a/2$ | 0 |
| $\alpha + u + 300° < \omega_o t < \alpha + 360°$ | 1, 2 | $e_a - e_c$ | 0 |

In the second subinterval ($\alpha + u < \omega_o t < \alpha + 60°$),

$$v_d = e_b - e_c \tag{3.14}$$

The average value of $v_d$ is

$$V_d = \frac{3}{\sqrt{2}\pi} V \left[\cos\alpha + \cos(\alpha + u)\right] \tag{3.15}$$

The harmonic components in the DC voltage are of order $h = 6k$, $k = 1, 2, 3, \ldots$
The rms value of the $h$th-order harmonic component is

$$V_h = \frac{3\sqrt{2}}{\pi} V \sqrt{\frac{\frac{\cos^2\left[(h-1)\frac{u}{2}\right]}{(h-1)^2} + \frac{\cos^2\left[(h+1)\frac{u}{2}\right]}{(h+1)^2} - \frac{2\cos\left[(h+1)\frac{u}{2}\right]\cos\left[(h-1)\frac{u}{2}\right]\cos(2\alpha+u)}{h^2-1}}{2}} \tag{3.16}$$

For each subinterval, the valves that conduct, DC voltage $v_d$, and voltage across valve 1 are given in Table 3.2.

The AC-side currents possess half-wave symmetry. The expression for one of these currents for half cycle is given by

$$i_b = \begin{cases} \frac{V}{\sqrt{2}\omega_o L}[\cos\alpha - \cos(\omega_o t)] & \text{if } \alpha \leq \omega_o t \leq \alpha + u \\ I_d & \text{if } \alpha + u \leq \omega_o t \leq \alpha + 120° \\ I_d - \frac{V}{\sqrt{2}\omega_o L}[\cos\alpha + \cos(\omega_o t + 60°)] & \text{if } \alpha + 120° \leq \omega_o t \leq \alpha + u + 120° \\ 0 & \text{if } \alpha + u + 120° \leq \omega_o t \leq \alpha + 180° \end{cases} \tag{3.17}$$

The rms value of the fundamental component of the AC-side currents is

$$I_1 = \frac{\sqrt{6}I_d}{2\pi}\sqrt{[\cos\alpha + \cos(\alpha+u)]^2 + \left[\frac{2u + \sin(2\alpha) - \sin(2\alpha + 2u)}{2\cos\alpha - 2\cos(\alpha + u)}\right]^2} \quad (3.18)$$

The harmonic components in the AC-side currents are of order $h = 6k \pm 1$, $k = 1, 2, 3, \ldots$ The rms value of the $h$th-order harmonic component is

$$I_h = \frac{\sqrt{6}I_d}{\pi h}\frac{\sqrt{\frac{\sin^2[(h+1)\frac{u}{2}]}{(h+1)^2} + \frac{\sin^2[(h-1)\frac{u}{2}]}{(h-1)^2} - \frac{2\sin[(h+1)\frac{u}{2}]\sin[(h-1)\frac{u}{2}]\cos(2\alpha+u)}{h^2-1}}}{\cos\alpha - \cos(\alpha+u)} \quad (3.19)$$

An angle known as extinction angle and denoted by $\gamma$ is defined [4] as

$$\gamma \triangleq \beta - u \quad (3.20)$$

For normal inverter operation, $120° < \alpha < 180°$, and the commutation margin angle is

$$\xi = \gamma \quad (3.21)$$

Equations (3.12) and (3.15) can be written in terms of $\gamma$ instead of $\alpha$ as follows:

$$V_d = \frac{3}{\sqrt{2}\pi}V\left[-\cos(\gamma + u) - \cos\gamma\right] \quad (3.22)$$

$$I_d = \frac{V}{\sqrt{2}\omega_o L}\left[-\cos(\gamma + u) + \cos\gamma\right] \quad (3.23)$$

### 3.2.1.3 Three- and Four-Valve Conduction Mode

Under certain abnormal conditions, $u > 60°$. If $u \geq 120°$, $v_d = 0$. For $60° < u < 120°$, one cycle of the AC voltage can be divided into six equal intervals and each interval can be divided into two subintervals. Four valves conduct in the first subinterval, and three valves conduct in the second subinterval. For steady-state analysis, it is sufficient to consider only one interval. The interval $\alpha < \omega_o t < \alpha + 60°$ is considered.

When valve 3 is turned on at $\omega_o t = \alpha$, valves 6, 1, and 2 are still conducting. The effective circuit to be analyzed in the first subinterval ($\alpha < \omega_o t < \alpha + u - 60°$) is shown in Fig. 3.6.

From the circuit diagram in Fig. 3.6,

**Fig. 3.6** Effective circuit
in the first subinterval for
three- and four-valve
conduction mode

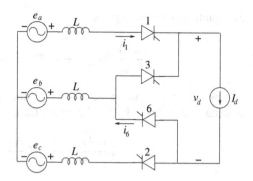

$$L\frac{d(I_d - i_1 - i_6)}{dt} - L\frac{di_1}{dt} = e_b - e_a \tag{3.24}$$

$$L\frac{d(I_d - i_1 - i_6)}{dt} + L\frac{d(I_d - i_6)}{dt} = e_b - e_c \tag{3.25}$$

Multiplying (3.25) by 2 and subtracting from (3.24) gives

$$3L\frac{di_6}{dt} = -\sqrt{6}V\cos(\omega_o t) \tag{3.26}$$

The solution of this equation is

$$i_6 = \frac{\sqrt{6}V}{3\omega_o L}[\sin\alpha - \sin(\omega_o t)] + i_6(\alpha) \tag{3.27}$$

Substituting $i_6(\alpha + u - 60°) = 0$ gives

$$i_6(\alpha) = \frac{\sqrt{6}V}{3\omega_o L}[\sin(\alpha + u - 60°) - \sin\alpha] \tag{3.28}$$

Substituting this in (3.27) gives

$$i_6 = \frac{\sqrt{6}V}{3\omega_o L}[\sin(\alpha + u - 60°) - \sin(\omega_o t)] \tag{3.29}$$

Multiplying (3.24) by 2 and subtracting from (3.25) give

$$3L\frac{di_1}{dt} = \sqrt{6}V\sin(\omega_o t + 150°) \tag{3.30}$$

The solution of this equation is

$$i_1 = \frac{\sqrt{6}V}{3\omega_o L}[\cos(\alpha + 150°) - \cos(\omega_o t + 150°)] + i_1(\alpha) \tag{3.31}$$

Substituting $i_1(\alpha) = I_d$ gives

$$i_1 = \frac{\sqrt{6}V}{3\omega_o L}[\cos(\alpha + 150°) - \cos(\omega_o t + 150°)] + I_d \qquad (3.32)$$

The effective circuit in the second subinterval $(\alpha + u - 60° < \omega_o t < \alpha + 60°)$ is same as that shown in Fig. 3.5. From the circuit diagram in Fig. 3.5,

$$L\frac{d(I_d - i_1)}{dt} - L\frac{di_1}{dt} = e_b - e_a \qquad (3.33)$$

The solution of this equation is

$$i_1 = \frac{V}{\sqrt{2}\omega_o L}[\cos(\omega_o t) - \cos(\alpha + u - 60°)] + i_1(\alpha + u - 60°) \qquad (3.34)$$

From (3.32),

$$i_1(\alpha + u - 60°) = \frac{\sqrt{6}V}{3\omega_o L}[\cos(\alpha + 150°) + \sin(\alpha + u)] + I_d \qquad (3.35)$$

Substituting this in (3.34) gives

$$i_1 = \frac{V}{\sqrt{2}\omega_o L}[\cos(\omega_o t) - \cos(\alpha + u - 60°)]$$
$$+ \frac{\sqrt{6}V}{3\omega_o L}[\cos(\alpha + 150°) + \sin(\alpha + u)] + I_d \qquad (3.36)$$

It is to be noted that

$$i_1(\alpha + 60°) = i_6(\alpha) \qquad (3.37)$$

Substituting for $i_1(\alpha + 60°)$ from (3.36) and for $i_6(\alpha)$ from (3.28) in (3.37) gives

$$I_d = \frac{V}{\sqrt{6}\omega_o L}[\cos(\alpha - 30°) - \cos(\alpha + u + 30°)] \qquad (3.38)$$

For each subinterval, the valves that conduct, DC voltage $v_d$, and voltage across valve 1 are given in Table 3.3.

The average value of $v_d$ is

$$V_d = \frac{3\sqrt{6}}{2\pi}V[\cos(\alpha - 30°) + \cos(\alpha + u + 30°)] \qquad (3.39)$$

**Table 3.3** DC-side voltage and voltage across a valve for three- and four-valve conduction mode

| Subinterval | Valves that conduct | $v_d$ | Voltage across valve 1 |
|---|---|---|---|
| $\alpha < \omega_o t < \alpha + u - 60°$ | 6, 1, 2, 3 | 0 | 0 |
| $\alpha + u - 60° < \omega_o t < \alpha + 60°$ | 1, 2, 3 | $-3e_c/2$ | 0 |
| $\alpha + 60° < \omega_o t < \alpha + u$ | 1, 2, 3, 4 | 0 | 0 |
| $\alpha + u < \omega_o t < \alpha + 120°$ | 2, 3, 4 | $3e_b/2$ | $-3e_b/2$ |
| $\alpha + 120° < \omega_o t < \alpha + u + 60°$ | 2, 3, 4, 5 | 0 | 0 |
| $\alpha + u + 60° < \omega_o t < \alpha + 180°$ | 3, 4, 5 | $-3e_a/2$ | $3e_a/2$ |
| $\alpha + 180° < \omega_o t < \alpha + u + 120°$ | 3, 4, 5, 6 | 0 | 0 |
| $\alpha + u + 120° < \omega_o t < \alpha + 240°$ | 4, 5, 6 | $3e_c/2$ | $-3e_c/2$ |
| $\alpha + 240° < \omega_o t < \alpha + u + 180°$ | 4, 5, 6, 1 | 0 | 0 |
| $\alpha + u + 180° < \omega_o t < \alpha + 300°$ | 5, 6, 1 | $-3e_b/2$ | 0 |
| $\alpha + 300° < \omega_o t < \alpha + u + 240°$ | 5, 6, 1, 2 | 0 | 0 |
| $\alpha + u + 240° < \omega_o t < \alpha + 360°$ | 6, 1, 2 | $3e_a/2$ | 0 |

Equations (3.38) and (3.39) can be written in terms of $\gamma$ instead of $\alpha$.

$$V_d = \frac{3\sqrt{6}}{2\pi} V \left[-\cos(\gamma + u + 30°) - \cos(\gamma - 30°)\right] \qquad (3.40)$$

$$I_d = \frac{V}{\sqrt{6}\omega_o L} \left[-\cos(\gamma + u + 30°) + \cos(\gamma - 30°)\right] \qquad (3.41)$$

### 3.2.2 12-Pulse Line-Commutated Converter

The circuit diagram of the 12-pulse converter is shown in Fig. 3.7 [4, 7]. The two six-pulse line-commutated converters are connected in series on the DC side to obtain a higher DC-side voltage. One transformer is wye–wye-connected, and the other is wye–delta-connected. The magnitude of the line-to-line voltages on the AC side of the two six-pulse converters should be same. Therefore, the number of turns in the transformer windings are as shown in Fig. 3.7.

If the transformers are assumed to be ideal, the three-phase voltage supplied by the delta-connected winding to one of the six-pulse converters lags the three-phase voltage supplied to the other six-pulse converter, by 30°. The instants of turn on of the thyristor valves of the six-pulse converter supplied by the delta-connected winding are delayed by 30° with respect to the instants of turn on of the thyristor valves of the other six-pulse converter. If $h$ is allowed to take a value 1 (for the fundamental component) or $6k \pm 1$, $k = 1, 2, 3, \ldots$ (for the harmonic components), the current phasors shown in Fig. 3.7 are given by

$$\mathbf{I}_{ah} = I_h\angle 0, \ \mathbf{I}_{bh} = I_h\angle(-120°h), \ \mathbf{I}_{ch} = I_h\angle(120°h) \qquad (3.42)$$

$$\mathbf{I}'_{ah} = I_h\angle(-30°h), \ \mathbf{I}'_{bh} = I_h\angle(-150°h), \ \mathbf{I}'_{ch} = I_h\angle(90°h) \qquad (3.43)$$

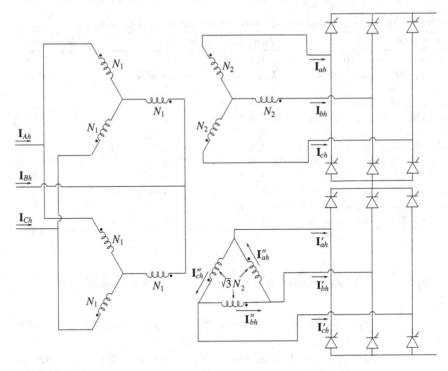

**Fig. 3.7** 12-pulse line-commutated converter

The currents $I''_{ah}$, $I''_{bh}$, and $I''_{ch}$ satisfy the following equations:

$$I''_{ah} - I''_{ch} = I'_{ah} \tag{3.44}$$

$$I''_{bh} - I''_{ah} = I'_{bh} \tag{3.45}$$

$$I''_{ah} + I''_{bh} + I''_{ch} = 0 \tag{3.46}$$

Equation (3.46) is obtained by applying Kirchhoff's current law to the neutral of the wye-connected winding. Solving (3.44)–(3.46) and using the relation $I'_{ah} + I'_{bh} + I'_{ch} = 0$ gives

$$I''_{ah} = \frac{1}{3}(I'_{ah} - I'_{bh}) \tag{3.47}$$

$$I''_{bh} = \frac{1}{3}(I'_{bh} - I'_{ch}) \tag{3.48}$$

$$I''_{ch} = \frac{1}{3}(I'_{ch} - I'_{ah}) \tag{3.49}$$

**Fig. 3.8** Schematic diagram
of a typical converter station

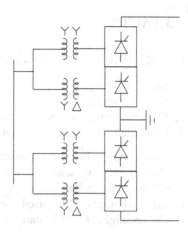

The currents in the AC system are

$$\mathbf{I}_{Ah} = \frac{N_2}{N_1}\mathbf{I}_{ah} + \sqrt{3}\frac{N_2}{N_1}\mathbf{I}''_{ah} = \begin{cases} 2\dfrac{N_2}{N_1}I_1\angle 0 & \text{if } h = 1 \\[2mm] 2\dfrac{N_2}{N_1}I_h\angle 0 & \text{if } h = 6k \pm 1, k = 2, 4, 6, \ldots \\[2mm] 0 & \text{if } h = 6k \pm 1, k = 1, 3, 5, \ldots \end{cases} \quad (3.50)$$

$$\mathbf{I}_{Bh} = \frac{N_2}{N_1}\mathbf{I}_{bh} + \sqrt{3}\frac{N_2}{N_1}\mathbf{I}''_{bh} = \begin{cases} 2\dfrac{N_2}{N_1}I_1\angle(-120°) & \text{if } h = 1 \\[2mm] 2\dfrac{N_2}{N_1}I_h\angle(\mp 120°) & \text{if } h = 6k \pm 1, k = 2, 4, 6, \ldots \\[2mm] 0 & \text{if } h = 6k \pm 1, k = 1, 3, 5, \ldots \end{cases} \quad (3.51)$$

$$\mathbf{I}_{Ch} = \frac{N_2}{N_1}\mathbf{I}_{ch} + \sqrt{3}\frac{N_2}{N_1}\mathbf{I}''_{ch} = \begin{cases} 2\dfrac{N_2}{N_1}I_1\angle 120° & \text{if } h = 1 \\[2mm] 2\dfrac{N_2}{N_1}I_h\angle(\pm 120°) & \text{if } h = 6k \pm 1, k = 2, 4, 6, \ldots \\[2mm] 0 & \text{if } h = 6k \pm 1, k = 1, 3, 5, \ldots \end{cases} \quad (3.52)$$

Therefore, the harmonic components in the AC system currents are of order
$h = 12k \pm 1, k = 1, 2, 3, \ldots$ The harmonic components in the DC-side voltage
are of order $h = 12k, k = 1, 2, 3, \ldots$

Typically, a converter station consists of two 12-pulse converters connected as
shown by the schematic diagram in Fig. 3.8 [4, 7]. One terminal on the DC side is at
a positive potential with respect to earth, and the other is at a negative potential. The
operation is such that the DC-side currents in the two 12-pulse converters are equal.

## 3.3 FACTS

FACTS controllers can be classified into the following types [3]:

1. Shunt controller.
2. Series controller.
3. Combined shunt–series controller.
4. Combined series–series controller.

Depending on the power semiconductor device used, the FACTS controllers can be classified into variable impedance-type controller and voltage source converter (VSC)-based controller. The device used in a variable impedance-type controller is thyristor, whereas a VSC-based controller uses a controllable switch.

The prominent FACTS controllers are as follows:

1. Static var compensator (SVC): variable impedance-type shunt controller.
2. Thyristor-controlled series capacitor (TCSC): variable impedance-type series controller.
3. Static synchronous compensator (STATCOM): VSC-based shunt controller.
4. Static synchronous series compensator (SSSC): VSC-based series controller.
5. Unified power flow controller (UPFC): VSC-based combined shunt–series controller.
6. Interline power flow controller (IPFC): VSC-based combined series–series controller.

The primary function of shunt FACTS controller is regulation of voltage and that of series FACTS controller is regulation of power flow.

### 3.3.1 SVC

The SVC consists of a three-phase thyristor-controlled reactor (TCR) in parallel with three capacitors connected in wye or delta.

#### 3.3.1.1 TCR

The TCR consists of a reactor and two thyristor valves connected in antiparallel. The effective reactance of TCR is varied by varying the instant of gate current pulses to the thyristor valves. Figure 3.9 shows a voltage source connected across a TCR where $v = \sqrt{2}V \cos(\omega_0 t)$ and $L$ is inductance.

If the gate current pulses are at the instants of voltage peaks, the effective reactance is minimum and is equal to that of the reactor. In order to increase the reactance, the gate current pulses to the thyristor valves are delayed. The gate current pulses to thyristor valves 1 and 2 are at $\omega_0 t = 2\pi k + \alpha$ and $\omega_0 t = 2\pi k + \alpha + \pi$,

**Fig. 3.9** TCR connected to a
voltage source

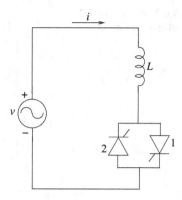

$k = 0, 1, 2, \ldots$, respectively, where $0 \le \alpha \le 90°$. The circuit of Fig. 3.9 is governed
by the following equation:

$$
\frac{di}{dt} =
\begin{cases}
\dfrac{\sqrt{2}V}{L} \cos(\omega_o t) & \text{if } \alpha < \omega_o t < \pi - \alpha \\
0 & \text{if } \pi - \alpha < \omega_o t < \pi + \alpha \\
\dfrac{\sqrt{2}V}{L} \cos(\omega_o t) & \text{if } \pi + \alpha < \omega_o t < 2\pi - \alpha \\
0 & \text{if } 2\pi - \alpha < \omega_o t < 2\pi + \alpha
\end{cases}
\tag{3.53}
$$

The solution of this equation is

$$
i =
\begin{cases}
\dfrac{\sqrt{2}V}{\omega_o L}[\sin(\omega_o t) - \sin \alpha] & \text{if } \alpha \le \omega_o t \le \pi - \alpha \\
0 & \text{if } \pi - \alpha \le \omega_o t \le \pi + \alpha \\
\dfrac{\sqrt{2}V}{\omega_o L}[\sin(\omega_o t) + \sin \alpha] & \text{if } \pi + \alpha \le \omega_o t \le 2\pi - \alpha \\
0 & \text{if } 2\pi - \alpha \le \omega_o t \le 2\pi + \alpha
\end{cases}
\tag{3.54}
$$

Figure 3.10 shows the plots of TCR voltage and current.
The rms value of the fundamental component of current is

$$
I_1 = \frac{V}{\pi \omega_o L} [\pi - 2\alpha - \sin(2\alpha)]
\tag{3.55}
$$

The current contains odd harmonic components only. The rms value of the odd
harmonic component of current, of order $h$, is

$$
I_h = \frac{4V |\sin \alpha \cos(h\alpha) - h \cos \alpha \sin(h\alpha)|}{\pi \omega_o L h(h^2 - 1)}
\tag{3.56}
$$

**Fig. 3.10** Plots of TCR
voltage and current

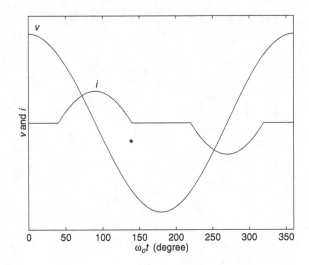

**Fig. 3.11** Six-pulse TCR

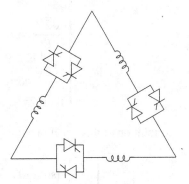

If the harmonic components are neglected, the susceptance of the TCR is

$$\frac{I_1}{V} = \frac{\pi - 2\alpha - \sin(2\alpha)}{\pi \omega_o L} \tag{3.57}$$

Connection of three TCRs in delta as shown in Fig. 3.11 results in the three-phase six-pulse TCR; the triplen harmonic components are eliminated in the line currents.

### 3.3.1.2 12-Pulse TCR

Half the number of harmonic components in the line currents of the six-pulse TCR can be eliminated using the 12-pulse TCR [3, 7]. Figure 3.12 shows the circuit

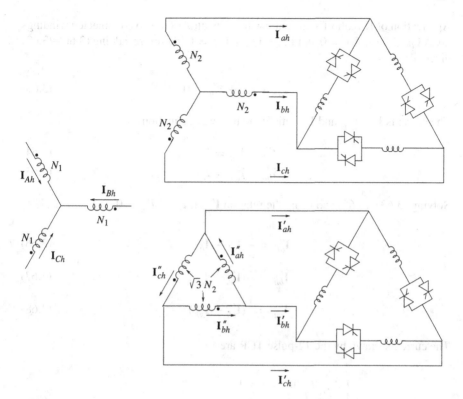

**Fig. 3.12** 12-pulse TCR

diagram of the 12-pulse TCR which consists of two six-pulse TCRs and a three-phase three-winding transformer.

   If transformers are assumed to be ideal, based on the explanation given in Sect. 3.2.2,

$$\mathbf{I}_{ah} = I_h\angle 0, \quad \mathbf{I}_{bh} = I_h\angle(-120°h), \quad \mathbf{I}_{ch} = I_h\angle(120°h) \tag{3.58}$$

$$\mathbf{I}'_{ah} = I_h\angle(-30°h), \quad \mathbf{I}'_{bh} = I_h\angle(-150°h), \quad \mathbf{I}'_{ch} = I_h\angle(90°h) \tag{3.59}$$

From (2.6),

$$\mathbf{I}_{Ah} = \frac{N_2}{N_1}\mathbf{I}_{ah} + \sqrt{3}\frac{N_2}{N_1}\mathbf{I}''_{ah} \tag{3.60}$$

$$\mathbf{I}_{Bh} = \frac{N_2}{N_1}\mathbf{I}_{bh} + \sqrt{3}\frac{N_2}{N_1}\mathbf{I}''_{bh} \tag{3.61}$$

$$\mathbf{I}_{Ch} = \frac{N_2}{N_1}\mathbf{I}_{ch} + \sqrt{3}\frac{N_2}{N_1}\mathbf{I}''_{ch} \tag{3.62}$$

Application of Kirchhoff's current law to the neutral of the wye-connected windings gives $I_{ah} + I_{bh} + I_{ch} = 0$ and $I_{Ah} + I_{Bh} + I_{Ch} = 0$; therefore, adding (3.60)–(3.62) gives

$$I''_{ah} + I''_{bh} + I''_{ch} = 0 \tag{3.63}$$

The currents $I''_{ah}$, $I''_{bh}$, and $I''_{ch}$ satisfy the following equations:

$$I''_{ah} - I''_{ch} = I'_{ah} \tag{3.64}$$

$$I''_{bh} - I''_{ah} = I'_{bh} \tag{3.65}$$

Solving (3.63)–(3.65) and using the relation $I'_{ah} + I'_{bh} + I'_{ch} = 0$ give

$$I''_{ah} = \frac{1}{3}(I'_{ah} - I'_{bh}) \tag{3.66}$$

$$I''_{bh} = \frac{1}{3}(I'_{bh} - I'_{ch}) \tag{3.67}$$

$$I''_{ch} = \frac{1}{3}(I'_{ch} - I'_{ah}) \tag{3.68}$$

The currents drawn by the 12-pulse TCR are

$$I_{Ah} = \begin{cases} 2\dfrac{N_2}{N_1} I_1 \angle 0 & \text{if } h = 1 \\ 2\dfrac{N_2}{N_1} I_h \angle 0 & \text{if } h = 6k \pm 1, k = 2, 4, 6, \ldots \\ 0 & \text{if } h = 6k \pm 1, k = 1, 3, 5, \ldots \end{cases} \tag{3.69}$$

$$I_{Bh} = \begin{cases} 2\dfrac{N_2}{N_1} I_1 \angle(-120°) & \text{if } h = 1 \\ 2\dfrac{N_2}{N_1} I_h \angle(\mp 120°) & \text{if } h = 6k \pm 1, k = 2, 4, 6, \ldots \\ 0 & \text{if } h = 6k \pm 1, k = 1, 3, 5, \ldots \end{cases} \tag{3.70}$$

$$I_{Ch} = \begin{cases} 2\dfrac{N_2}{N_1} I_1 \angle 120° & \text{if } h = 1 \\ 2\dfrac{N_2}{N_1} I_h \angle(\pm 120°) & \text{if } h = 6k \pm 1, k = 2, 4, 6, \ldots \\ 0 & \text{if } h = 6k \pm 1, k = 1, 3, 5, \ldots \end{cases} \tag{3.71}$$

Therefore, the harmonic components in the currents drawn by a 12-pulse TCR are of order $h = 12k \pm 1, k = 1, 2, 3, \ldots$

**Fig. 3.13** SVC and the
Thevenin equivalent circuit of
the network

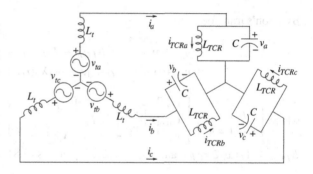

### 3.3.1.3 Controller for SVC

In steady state, if the harmonic components in the TCR currents are neglected, the
TCR can be represented by three wye-connected inductors of inductance $L_{TCR}$. Let
the capacitors of capacitance $C$ be wye-connected. SVC is connected at a bus in a
network. The network is represented by the Thevenin equivalent circuit shown in
Fig. 3.13. The resistance in the Thevenin equivalent circuit is neglected, and $L_t$ is
inductance.

The equations governing the circuit in Fig. 3.13 are

$$L_t \frac{di_a}{dt} = v_{ta} - v_a \tag{3.72}$$

$$L_t \frac{di_b}{dt} = v_{tb} - v_b \tag{3.73}$$

$$L_t \frac{di_c}{dt} = v_{tc} - v_c \tag{3.74}$$

$$L_{TCR} \frac{di_{TCRa}}{dt} = v_a \tag{3.75}$$

$$L_{TCR} \frac{di_{TCRb}}{dt} = v_b \tag{3.76}$$

$$L_{TCR} \frac{di_{TCRc}}{dt} = v_c \tag{3.77}$$

$$C \frac{dv_a}{dt} = i_a - i_{TCRa} \tag{3.78}$$

$$C \frac{dv_b}{dt} = i_b - i_{TCRb} \tag{3.79}$$

$$C \frac{dv_c}{dt} = i_c - i_{TCRc} \tag{3.80}$$

By Kron's transformation,

$$i_Q + ji_D = -j(B_{TCR} - B_C)(v_Q + jv_D) \tag{3.81}$$

$$v_Q + jv_D = v_{tQ} + jv_{tD} - jX_t(i_Q + ji_D) \tag{3.82}$$

where $\begin{bmatrix} v_{tD} \\ v_{tQ} \end{bmatrix} \triangleq T'_K \begin{bmatrix} v_{ta} \\ v_{tb} \\ v_{tc} \end{bmatrix}$, $\begin{bmatrix} v_D \\ v_Q \end{bmatrix} \triangleq T'_K \begin{bmatrix} v_a \\ v_b \\ v_c \end{bmatrix}$, $\begin{bmatrix} i_D \\ i_Q \end{bmatrix} \triangleq T'_K \begin{bmatrix} i_a \\ i_b \\ i_c \end{bmatrix}$, $B_C \triangleq \omega_o C$,

$B_{TCR} \triangleq 1/(\omega_o L_{TCR})$, and $X_t \triangleq \omega_o L_t$. Equations (3.81) and (3.82) can be written as

$$I\angle(\phi \pm 90°) = -jB_{SVC}V\angle\phi \tag{3.83}$$

$$V\angle\phi = V_t\angle\phi - jI\angle(\phi \pm 90°)X_t \tag{3.84}$$

where $V \triangleq \sqrt{v_D^2 + v_Q^2}$, $I \triangleq \sqrt{i_D^2 + i_Q^2}$, $\phi \triangleq \tan^{-1}(v_D/v_Q) = \tan^{-1}(v_{tD}/v_{tQ})$, $V_t \triangleq \sqrt{v_{tD}^2 + v_{tQ}^2}$, and $B_{SVC} \triangleq B_{TCR} - B_C$. Equations (3.83) and (3.84) can be written as

$$i_R = B_{SVC}V \tag{3.85}$$

$$V = V_t - i_R X_t \tag{3.86}$$

where

$$i_R = \begin{cases} I & \text{if current is inductive} \\ -I & \text{if current is capacitive} \end{cases} \tag{3.87}$$

$i_R$ is reactive current as defined by (2.135). For a given value of $B_{SVC}$, (3.85) is the equation of a straight line in the $i_R$–$V$ plane. A few representative straight lines corresponding to different values of $B_{SVC}$ are shown in Fig. 3.14. Let $0 \leq B_{TCR} \leq B_L$; the minimum value of $B_{SVC}$ is $-B_C$, and the maximum value of $B_{SVC}$ is $B_L - B_C$. Equation (3.86) is that of a straight line shown in Fig. 3.15.

For a given steady-state operating condition, the values of $V$ and $i_R$ are obtained by solving (3.85) and (3.86). In order to regulate the voltage, a regulator is used to vary the value of $B_{SVC}$ such that in steady state, the voltage magnitude $V$ is equal to the desired value $V_{ref}$. The block diagram of the regulator is shown in Fig. 3.16 where the regulator is typically a proportional–integral controller. The proportional and integral gains are negative; the sign of the proportional and integral gains should be such that the change in the regulator output results in reduction in the magnitude of the error (regulator input). $B_{ref}$ is the desired value of $B_{SVC}$. In steady state, the $V$ versus $i_R$ characteristic is shown in Fig. 3.17. It consists of three straight lines: One straight line corresponds to the inductive limit at which $B_{SVC} = B_L - B_C$, one straight line corresponds to the capacitive limit at which $B_{SVC} = -B_C$, and the straight line with zero slope corresponds to the control range.

**Fig. 3.14** Plot of voltage
magnitude versus reactive
current for different values of
SVC susceptance

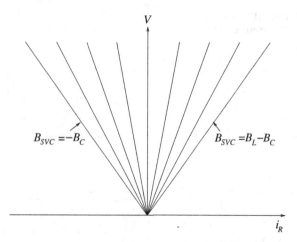

$B_{SVC} = -B_C$        $B_{SVC} = B_L - B_C$

**Fig. 3.15** Characteristic of
the network

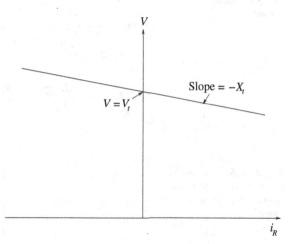

$V = V_t$        Slope $= -X_t$

**Fig. 3.16** SVC controller

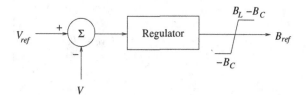

$V_{ref}$        $\Sigma$        Regulator        $B_L - B_C$        $B_{ref}$

$-B_C$

$V$

As the power system operating condition changes, the values of $V_t$ and $X_t$ change,
and this may result in operation at the inductive or capacitive limits. To avoid fre-
quent hitting of the limits, the steady-state characteristic is altered using the controller
shown in Fig. 3.18 [3]; $X_s$ is positive. Due to this controller, the steady-state characteristic is as shown in Fig. 3.19.

**Fig. 3.17** Steady-state characteristic of SVC

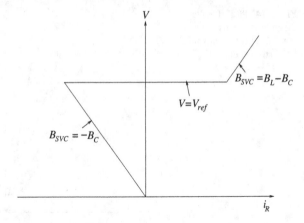

**Fig. 3.18** Modified SVC controller

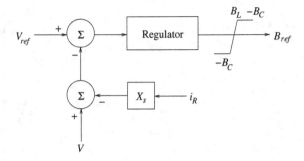

**Fig. 3.19** Steady-state characteristic of SVC with positive slope in the control range

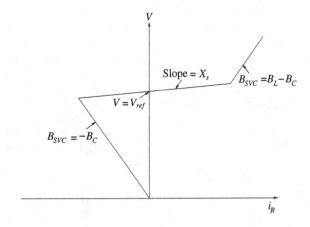

### 3.3.2 TCSC

A TCSC is connected in series with a transmission line. The TCSC consists of a single-phase TCR and a capacitor in parallel, in all the three phases.

**Fig. 3.20** TCSC connected in series with a current source

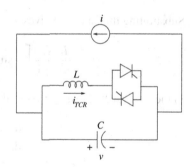

### 3.3.2.1 Analysis of TCSC

The effective reactance of TCSC is varied by varying the instant of gate current pulses to the thyristors. Figure 3.20 shows one phase of a TCSC where $L$ is inductance and $C$ is capacitance; the transmission line current is assumed to be independent and is modelled by a current source $i = I_m \cos(\omega_o t)$ [3, 8].

If the thyristors are turned on at the instants of peaks of current $i$, the current through the TCR is zero. In order to vary the effective reactance of the TCSC, the gate current pulses to the thyristors are advanced by an angle $\beta$ with respect to the instants of peaks of current $i$; $0 \le \beta \le \pi/2$. The equations governing the circuit, when the TCR conducts, are

$$\frac{dv}{dt} = -\frac{i_{TCR}}{C} + \frac{I_m \cos \omega_o t}{C} \tag{3.88}$$

$$\frac{di_{TCR}}{dt} = \frac{v}{L} \tag{3.89}$$

In steady state, $i_{TCR}$ and $v$ possess quarter-wave symmetry; the instant at which $i_{TCR}$ attains the peak value and the instant of zero crossing of $v$ are the same as the instant at which $i$ attains the peak value. The steady-state solution of (3.88) and (3.89), in the interval $-\beta \le \omega_o t \le \beta$, is given by the following two equations:

$$v = \frac{I_m X_C}{2} \left[ \frac{\sin(\lambda\omega_o t + \lambda\beta + \beta) + \sin(\omega_o t)}{\lambda + 1} + \frac{\sin(\lambda\omega_o t + \lambda\beta - \beta) - \sin(\omega_o t)}{\lambda - 1} \right]$$

$$+ v(-\beta) \cos(\lambda\omega_o t + \lambda\beta) \tag{3.90}$$

$$i_{TCR} = \frac{I_m \lambda}{2} \left[ \frac{\cos(\omega_o t) - \cos(\lambda\omega_o t + \lambda\beta + \beta)}{\lambda + 1} \right.$$

$$\left. + \frac{\cos(\omega_o t) - \cos(\lambda\omega_o t + \lambda\beta - \beta)}{\lambda - 1} \right] + \sqrt{\frac{C}{L}} v(-\beta) \sin(\lambda\omega_o t + \lambda\beta) \tag{3.91}$$

where $\lambda \triangleq 1/(\omega_o\sqrt{LC})$ and $X_C \triangleq 1/(\omega_o C)$. Substituting $v(0) = 0$ in (3.90) gives

$$v(-\beta) = \frac{I_m X_C}{\lambda^2 - 1} [\sin \beta - \lambda \cos \beta \tan(\lambda\beta)] \tag{3.92}$$

Substituting this in (3.90) gives

$$v = \frac{I_m X_C}{\lambda^2 - 1} \left[ -\sin(\omega_o t) + \frac{\lambda \cos \beta \sin(\lambda \omega_o t)}{\cos(\lambda \beta)} \right] \tag{3.93}$$

In the interval $\beta \leq \omega_o t \leq \pi - \beta$, the equation governing the circuit is

$$\frac{dv}{dt} = \frac{I_m \cos(\omega_o t)}{C} \tag{3.94}$$

$v(\beta) = -v(-\beta)$. Therefore, the solution of (3.94) is

$$v = \frac{I_m X_C}{\lambda^2 - 1} [\lambda \cos \beta \tan(\lambda \beta) - \sin \beta] + I_m X_C [\sin(\omega_o t) - \sin \beta] \tag{3.95}$$

The peak value of the fundamental component of $v$ is

$$V_{1m} = \frac{4}{\pi} \int_0^{\pi/4} v \sin(\omega_o t) d(\omega_o t) \tag{3.96}$$

If the harmonic components are neglected, the reactance of TCSC is

$$X_{TCSC} = \frac{V_{1m}}{I_m} = X_C + \frac{2}{\pi} \frac{\lambda^2 X_C}{\lambda^2 - 1} \left[ -\frac{1}{2} \frac{\lambda^2 + 1}{\lambda^2 - 1} \sin(2\beta) - \beta + \frac{2\lambda \cos^2 \beta \tan(\lambda \beta)}{\lambda^2 - 1} \right] \tag{3.97}$$

$X_{TCSC}$ is positive when it is capacitive and negative when it is inductive. Parallel resonance occurs at certain values of $\beta$, denoted by $\beta_{res}$. $X_{TCSC} \to \infty$ as $\beta \to \beta_{res} = (2k+1)\pi/(2\lambda)$, $k = 0, 1, 2, \ldots$ Since $0 \leq \beta \leq \pi/2$, there is only one value of $\beta_{res}$ if $\lambda < 3$. Typically, $\lambda < 3$ [3, 8], and hence, $\beta_{res} = \pi/(2\lambda)$. Figure 3.21 shows the variation of $X_{TCSC}$ with $\beta$.

### 3.3.2.2  Controller for TCSC

The primary purpose of TCSC is to increase the power flow in a transmission line. Since the reactance of TCSC can be varied, it can be controlled in order to regulate the power flow in the transmission line. Normally, $X_{TCSC}$ is capacitive. Operation near resonance results in a large voltage across the TCSC. To limit the voltage across TCSC, $0 \leq \beta \leq \beta_{max}$ where $\beta_{max} < \beta_{res}$. At $\beta = 0$, $X_{TCSC} = X_C$, and at $\beta = \beta_{max}$, $X_{TCSC} = X_{max}$.

The most basic type of control is open-loop control where $\beta$ is decided based on the desired TCSC reactance $X_{ref}$. The other type of control is closed-loop control. One type of closed-loop control is the constant current (CC) control [3, 8]. In CC control,

**Fig. 3.21** Plot of TCSC reactance as a function of angle $\beta$

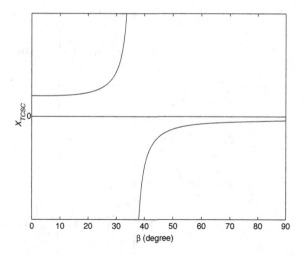

**Fig. 3.22** CC controller for TCSC

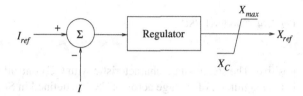

**Fig. 3.23** Steady-state characteristic of TCSC with CC controller

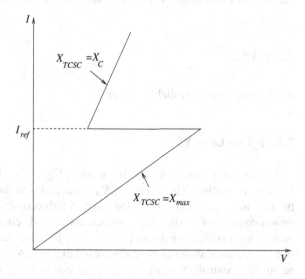

the magnitude of the current in the transmission line is regulated. The block diagram of CC controller is shown in Fig. 3.22. $I$ is the magnitude of the current through TCSC as defined in Sect. 2.3.1, and $I_{ref}$ is the desired value of $I$. The regulator is typically a proportional–integral controller; the proportional and integral gains are

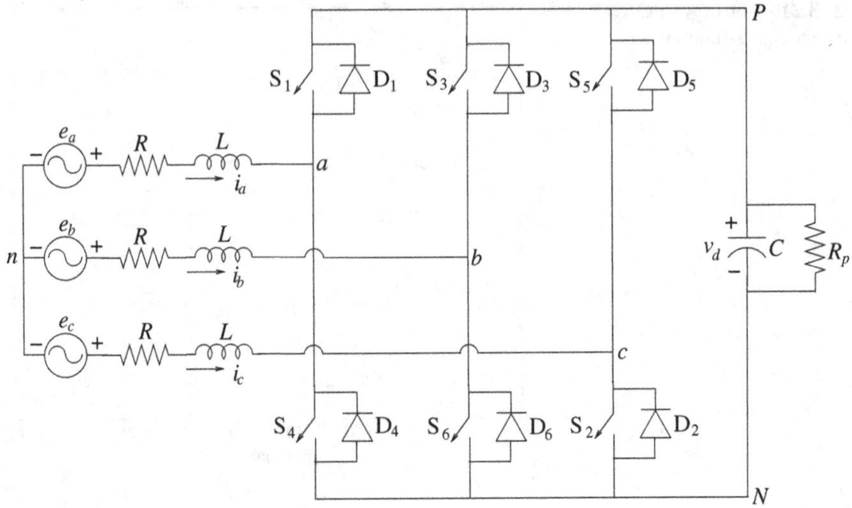

**Fig. 3.24** Two-level VSC

positive. The steady-state characteristic with CC controller is shown in Fig. 3.23. $V$
is the magnitude of voltage across TCSC as defined in Sect. 2.3.1. In steady state, if
the harmonic components are neglected, $V$ and $I$ are related by $V = X_{TCSC} I$.

### 3.3.3 VSC

A VSC employs controllable switches.

#### 3.3.3.1  Two-Level VSC

The circuit of a two-level VSC is shown in Fig. 3.24.  The VSC consists of three
legs and a capacitor. The resistance $R_p$ in parallel with the capacitance $C$ represents
the loss in the capacitor. Each leg has two bidirectional switches. A bidirectional
switch consists of a controllable switch and a diode connected in antiparallel. The
bidirectional switch consisting of $S_1$ and $D_1$ is called switch 1. Similarly, the other
five bidirectional switches are called switches 2 to 6. A bidirectional switch is said to
be on if the controllable switch or the diode conducts. At any instant, in each leg, one
bidirectional switch is on and one bidirectional switch is off. The DC-side voltage
$v_d$ is always positive. The possible states of one of the legs are shown in Table 3.4.
$v_{aN}$ is the voltage of terminal $a$ with respect to the DC-side negative terminal $N$. The
circuit on the AC side is represented by a three-phase balanced voltage source with
a resistance $R$ and an inductance $L$ in each phase. The currents $i_a$, $i_b$, and $i_c$ flow

**Table 3.4** Possible states of a leg of the two-level VSC

| Controllable switch with control signal to turn on | Controllable switch with control signal to turn off | $v_{aN}$ | Device that conducts $i_a > 0$ | Device that conducts $i_a < 0$ |
|---|---|---|---|---|
| $S_1$ | $S_4$ | $v_d$ | $D_1$ | $S_1$ |
| $S_4$ | $S_1$ | 0 | $S_4$ | $D_4$ |

through inductances and hence can vary only at a finite rate. If $D_1$ is conducting and a control signal is given to turn on $S_4$, then $S_4$ starts conducting and the capacitor voltage reverse-biases $D_1$; if $S_1$ is conducting and it is turned off, then $D_4$ starts conducting. Similarly, if $D_4$ is conducting and a control signal is given to turn on $S_1$, then $S_1$ starts conducting and the capacitor voltage reverse-biases $D_4$; if $S_4$ is conducting and it is turned off, then $D_1$ starts conducting.

The adjective 'two-level' means that for a given DC-side voltage, the number of possible values of the voltage of an AC-side terminal with respect to a DC-side terminal is two. For example, $v_{aN}$ is equal to either $v_d$ or 0.

Let $v_{an}$, $v_{bn}$, and $v_{cn}$ be the voltages of the terminals $a$, $b$, and $c$, respectively, with respect to the neutral $n$ of the AC side. Let $v_{bN}$ and $v_{cN}$ be the voltages of the terminals $b$ and $c$, respectively, with respect to the terminal $N$. Let $v_{Nn}$ be the voltage of $N$ with respect to $n$.

$$v_{an} = v_{aN} + v_{Nn} \tag{3.98}$$

$$v_{bn} = v_{bN} + v_{Nn} \tag{3.99}$$

$$v_{cn} = v_{cN} + v_{Nn} \tag{3.100}$$

The switches are turned on and off such that the fundamental components of AC-side currents $i_a$, $i_b$, and $i_c$ are equal in magnitude and displaced by 120°. The triplen harmonic components of these currents are zero. For any other harmonic order, the harmonic components of the three currents are equal in magnitude and displaced by 120°. Hence, the drops across the series combination of $R$ and $L$ in the three phases add to zero. Therefore, since the AC-side source voltages, $e_a$, $e_b$, and $e_c$ are balanced,

$$v_{an} + v_{bn} + v_{cn} = 0 \tag{3.101}$$

Adding (3.98)–(3.100) and using (3.101) give

$$v_{Nn} = -\frac{1}{3}(v_{aN} + v_{bN} + v_{cN}) \tag{3.102}$$

Hence, from (3.98) to (3.100) and (3.102),

$$\begin{bmatrix} v_{an} \\ v_{bn} \\ v_{cn} \end{bmatrix} = \begin{bmatrix} 2/3 & -1/3 & -1/3 \\ -1/3 & 2/3 & -1/3 \\ -1/3 & -1/3 & 2/3 \end{bmatrix} \begin{bmatrix} v_{aN} \\ v_{bN} \\ v_{cN} \end{bmatrix} \tag{3.103}$$

The voltages $v_{aN}$, $v_{bN}$, and $v_{cN}$ can be written in terms of switching functions $u_a$, $u_b$, and $u_c$.

$$
\begin{bmatrix} v_{aN} \\ v_{bN} \\ v_{cN} \end{bmatrix} = \begin{bmatrix} u_a \\ u_b \\ u_c \end{bmatrix} v_d \tag{3.104}
$$

where $u_a = 1$ if switch 1 is on and $u_a = 0$ if switch 4 in on, $u_b = 1$ if switch 3 is on and $u_b = 0$ if switch 6 in on, and $u_c = 1$ if switch 5 is on and $u_c = 0$ if switch 2 in on. From (3.103) and (3.104),

$$
\begin{bmatrix} v_{an} \\ v_{bn} \\ v_{cn} \end{bmatrix} = \begin{bmatrix} s_a \\ s_b \\ s_c \end{bmatrix} v_d \tag{3.105}
$$

where

$$
\begin{bmatrix} s_a \\ s_b \\ s_c \end{bmatrix} \triangleq \begin{bmatrix} 2/3 & -1/3 & -1/3 \\ -1/3 & 2/3 & -1/3 \\ -1/3 & -1/3 & 2/3 \end{bmatrix} \begin{bmatrix} u_a \\ u_b \\ u_c \end{bmatrix} \tag{3.106}
$$

$s_a$, $s_b$, and $s_c$ are also called switching functions.

The switching of the GTO thyristor is slow. If GTO thyristors are used as controllable switches, in order to minimize the switching losses, each GTO thyristor is turned on and off only once in a cycle [3]. The plots of $u_a$, $u_b$, and $u_c$ are shown in Fig. 3.25; the plots of $s_a$, $s_b$, and $s_c$ are shown in Fig. 3.26. With this switching, for a given constant $v_d$, the magnitude of the fundamental component of $v_{an}$, $v_{bn}$, and $v_{cn}$ is fixed. If $v_d$ is constant, the harmonic components present in $v_{an}$, $v_{bn}$, and $v_{cn}$ are of order $h = 6k \pm 1, k = 1, 2, 3, \ldots$

If IGBT is used as the controllable switch, the switching frequency can be increased. Then, the magnitude of the fundamental component of the voltages $v_{an}$, $v_{bn}$, and $v_{cn}$ can be varied for a given constant $v_d$ using the switching function $u_a'$ shown in Fig. 3.27 instead of $u_a$. This switching function has a notch in both positive and negative half cycles and possesses quarter-wave symmetry. In addition to controlling the magnitude of the fundamental component, some harmonic components can be eliminated by introducing more notches in the switching function. For example, using the switching function $u_a''$ shown in Fig. 3.28 which has two notches in both positive and negative half cycles, a harmonic component can be eliminated in addition to controlling the magnitude of the fundamental component; there are two degrees of freedom, namely $\beta$ and $\gamma$. $u_a''$ should possess quarter-wave symmetry. This type of switching is called selective harmonic elimination.

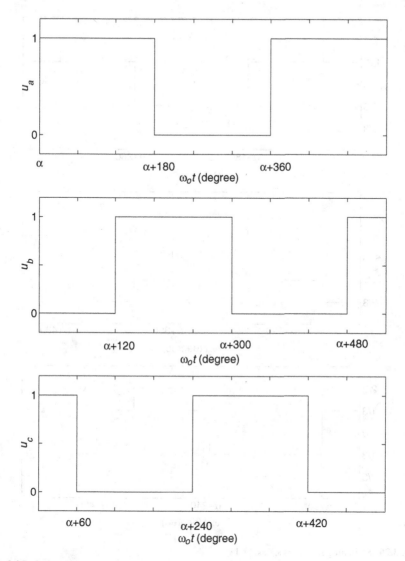

**Fig. 3.25** Switching functions $u_a$, $u_b$, and $u_c$

### 3.3.3.2 Multi-Level VSC

Diode-clamped converter is a type of multi-level VSC. Figure 3.29 shows the circuit diagram of a three-level diode-clamped converter. In addition to the bidirectional switches, there are six diodes $D_{c1}$ to $D_{c6}$ which are called clamping diodes. The resistances $R_p$ in parallel with the capacitances $C$ represent the loss in the capacitors.

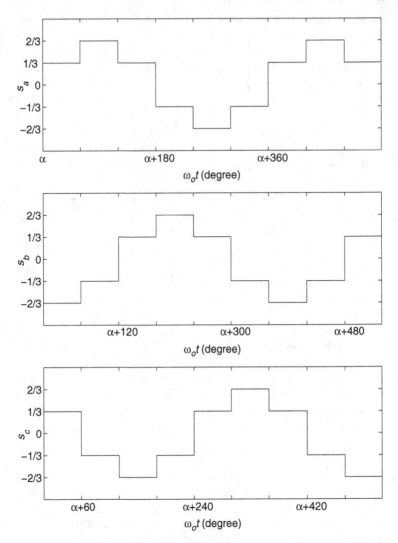

**Fig. 3.26** Switching functions $s_a$, $s_b$, and $s_c$

It is assumed that $v_{d1} \approx v_{d2} \approx v_d/2$. $v_d$ is always positive. For each leg, there are three possible states. The possible states of one of the legs are shown in Table 3.5. $v_{am}$ is the voltage of the terminal $a$ with respect to the midpoint $m$ of the DC side.

Let $v_{an}$, $v_{bn}$, and $v_{cn}$ be the voltages of the terminals $a$, $b$, and $c$, respectively, with respect to the neutral $n$ of the AC side. Let $v_{bm}$ and $v_{cm}$ be the voltages of the terminals $b$ and $c$, respectively, with respect to the midpoint $m$ of the DC side. $v_{am}$, $v_{bm}$, and $v_{cm}$ can be written in terms of switching functions $u_{3a}$, $u_{3b}$, and $u_{3c}$.

**Fig. 3.27** Switching function $u'_a$

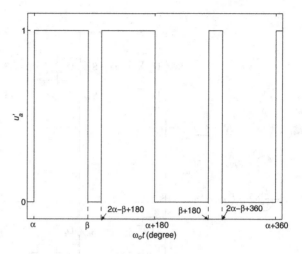

**Fig. 3.28** Switching function $u''_a$

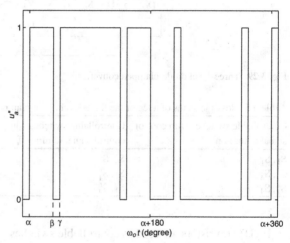

$$v_{am} = u_{3a}v_d \qquad (3.107)$$
$$v_{bm} = u_{3b}v_d \qquad (3.108)$$
$$v_{cm} = u_{3c}v_d \qquad (3.109)$$

where $u_{3a}$, $u_{3b}$, and $u_{3c}$ are equal to $1/2$, $-1/2$, or $0$, depending on the status of the switches. For a given value of $v_d$, the number of possible values of $v_{am}$, $v_{bm}$, and $v_{cm}$ is three; hence, this VSC is called three-level VSC. Similar to the derivation of (3.103), it can be shown that

$$\begin{bmatrix} v_{an} \\ v_{bn} \\ v_{cn} \end{bmatrix} = \begin{bmatrix} 2/3 & -1/3 & -1/3 \\ -1/3 & 2/3 & -1/3 \\ -1/3 & -1/3 & 2/3 \end{bmatrix} \begin{bmatrix} v_{am} \\ v_{bm} \\ v_{cm} \end{bmatrix} \qquad (3.110)$$

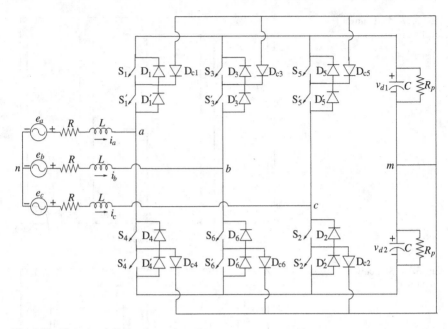

**Fig. 3.29**  Three-level diode-clamped converter

**Table 3.5**  Possible states of a leg of the three-level diode-clamped VSC

| Controllable switches with control signal to turn on | Controllable switches with control signal to turn off | $v_{am}$ | Devices that conduct | |
|---|---|---|---|---|
| | | | $i_a > 0$ | $i_a < 0$ |
| $S_1, S_1'$ | $S_4, S_4'$ | $v_d/2$ | $D_1', D_1$ | $S_1, S_1'$ |
| $S_4, S_4'$ | $S_1, S_1'$ | $-v_d/2$ | $S_4, S_4'$ | $D_4', D_4$ |
| $S_1', S_4$ | $S_1, S_4'$ | 0 | $S_4, D_{c4}$ | $D_{c1}, S_1'$ |

If GTO thyristors are used as controllable switches, each GTO thyristor is turned on and off only once in a cycle and $u_{3a}$ is as shown in Fig. 3.30. $u_{3a}$ possesses quarter-wave symmetry. The magnitude of the fundamental component of $v_{an}$, $v_{bn}$, and $v_{cn}$ is controlled by varying $\beta$. If $v_d$ is constant, the harmonic components present in $v_{an}$, $v_{bn}$, and $v_{cn}$ are of order $h = 6k \pm 1, k = 1, 2, 3, \ldots$ If IGBT is used as the controllable switch, in addition to controlling the magnitude of the fundamental component, some harmonic components can be eliminated by increasing the switching frequency.

### 3.3.3.3  Multi-Pulse VSC

The VSCs described in Sects. 3.3.3.1 and 3.3.3.2 are called six-pulse VSCs. A multi-pulse VSC is used to reduce the harmonic content in the AC-side voltages. The circuit diagram of the 12-pulse VSC, which is a multi-pulse VSC, is shown in Fig. 3.31 [3].

**Fig. 3.30** Switching functions $u_{3a}$

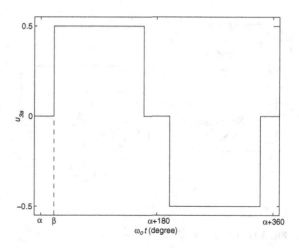

The subscript $h$ takes a value of 1 for the fundamental component and $6k \pm 1$, $k = 1, 2, 3, \ldots$ for the harmonic components. Let $\mathbf{V}_{ah}$, $\mathbf{V}_{bh}$, and $\mathbf{V}_{ch}$ be the voltages of terminals $a$, $b$, and $c$, respectively, with respect to the neutral. Let $\mathbf{V}'_{ah}$, $\mathbf{V}'_{bh}$, and $\mathbf{V}'_{ch}$ be the voltages of terminals $a'$, $b'$, and $c'$, respectively, with respect to the neutral. The control signals to the controllable switches of the two six-pulse VSCs are given such that

$$\mathbf{V}_{a1} = V_1 \angle 0, \quad \mathbf{V}_{b1} = V_1 \angle(-120°), \quad \mathbf{V}_{c1} = V_1 \angle 120° \qquad (3.111)$$

$$\mathbf{V}'_{a1} = V_1 \angle(-30°), \quad \mathbf{V}'_{b1} = V_1 \angle(-150°), \quad \mathbf{V}'_{c1} = V_1 \angle 90° \qquad (3.112)$$

The general expressions for the voltage phasors applicable to both the fundamental component and the harmonic components are

$$\mathbf{V}_{ah} = V_h \angle 0, \quad \mathbf{V}_{bh} = V_h \angle(-120°h), \quad \mathbf{V}_{ch} = V_h \angle(120°h) \qquad (3.113)$$

$$\mathbf{V}'_{ah} = V_h \angle(-30°h), \quad \mathbf{V}'_{bh} = V_h \angle(-150°h), \quad \mathbf{V}'_{ch} = V_h \angle(90°h) \qquad (3.114)$$

If the transformer is assumed to be ideal, from Fig. 3.31, the emfs in the transformer windings are given by

$$\mathbf{V}_{Ah} = \mathbf{V}'_{ah} - \mathbf{V}'_{bh}, \quad \mathbf{V}_{Bh} = \mathbf{V}'_{bh} - \mathbf{V}'_{ch}, \quad \mathbf{V}_{Ch} = \mathbf{V}'_{ch} - \mathbf{V}'_{ah} \qquad (3.115)$$

$$\mathbf{V}'_{Ah} = \frac{1}{\sqrt{3}} \mathbf{V}_{Ah}, \quad \mathbf{V}'_{Bh} = \frac{1}{\sqrt{3}} \mathbf{V}_{Bh}, \quad \mathbf{V}'_{Ch} = \frac{1}{\sqrt{3}} \mathbf{V}_{Ch} \qquad (3.116)$$

The voltages at the terminals $a''$, $b''$, and $c''$ with respect to the neutral are

$$\mathbf{V}''_{ah} = \mathbf{V}'_{Ah} + \mathbf{V}_{ah} = \begin{cases} 2V_1 \angle 0 & \text{if } h = 1 \\ 2V_h \angle 0 & \text{if } h = 6k \pm 1, k = 2, 4, 6, \ldots \\ 0 & \text{if } h = 6k \pm 1, k = 1, 3, 5, \ldots \end{cases} \qquad (3.117)$$

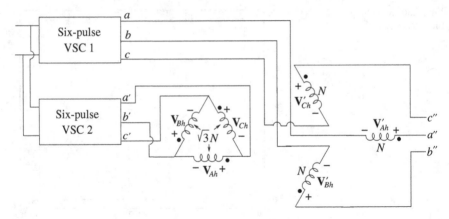

**Fig. 3.31** 12-pulse VSC

$$\mathbf{V}''_{bh} = \mathbf{V}'_{Bh} + \mathbf{V}_{bh} = \begin{cases} 2V_1\angle(-120°) & \text{if } h = 1 \\ 2V_h\angle(\mp120°) & \text{if } h = 6k \pm 1, k = 2, 4, 6, \dots \\ 0 & \text{if } h = 6k \pm 1, k = 1, 3, 5, \dots \end{cases} \quad (3.118)$$

$$\mathbf{V}''_{ch} = \mathbf{V}'_{Ch} + \mathbf{V}_{ch} = \begin{cases} 2V_1\angle120° & \text{if } h = 1 \\ 2V_h\angle(\pm120°) & \text{if } h = 6k \pm 1, k = 2, 4, 6, \dots \\ 0 & \text{if } h = 6k \pm 1, k = 1, 3, 5, \dots \end{cases} \quad (3.119)$$

Therefore, the order of the harmonic components in the AC-side voltages of the 12-pulse VSC is $h = 12k \pm 1$, $k = 1, 2, 3, \dots$

The concept of harmonic cancellation can be generalized. The schematic diagram of a general multi-pulse VSC is shown in Fig. 3.32. It consists of $n$ identical six-pulse VSCs which are connected in parallel on the DC side and in series on the AC side through transformers. Each transformer has three windings in each phase. One winding is wye-connected, and the other two windings are connected in zigzag. The wye-connected winding is connected to the VSC. The circuit diagram of transformer $i$ is shown in Fig. 3.33. The subscript $h$ is equal to 1 for the fundamental component and is equal to $6k \pm 1$, $k = 1, 2, 3, \dots$ for the harmonic components.

The control signals to the controllable switches of the six-pulse VSCs are given such that the fundamental components of the voltages, with respect to the neutral, at the AC-side terminals of $(i + 1)$th VSC are $V_1\angle(-60°i/n)$, $V_1\angle(-60°i/n - 120°)$, and $V_1\angle(-60°i/n + 120°)$, $i = 0, 1, 2, \dots n - 1$ [9]. For $i = 1, 2, 3, \dots n - 1$, the voltages at the terminals of the $(i + 1)$th VSC are nothing but the voltages across the wye-connected winding of transformer $i$ shown in Fig. 3.33. Therefore,

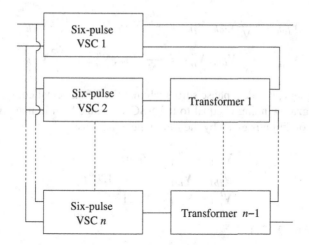

**Fig. 3.32** Schematic diagram of multi-pulse VSC

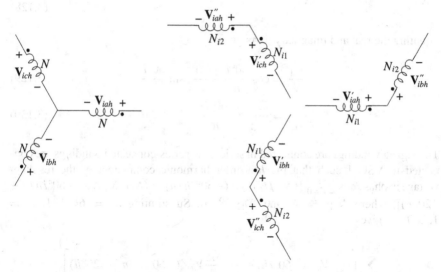

**Fig. 3.33** Transformer $i$ of multi-pulse VSC

$$\mathbf{V}_{ia1} = V_1\angle(-60°i/n) \tag{3.120}$$

$$\mathbf{V}_{ib1} = V_1\angle(-60°i/n - 120°) \tag{3.121}$$

$$\mathbf{V}_{ic1} = V_1\angle(-60°i/n + 120°) \tag{3.122}$$

If it is assumed that the transformers are ideal, the other voltages shown in Fig. 3.33 are given by

$$\mathbf{V}'_{iah} = \frac{N_{i1}}{N}\mathbf{V}_{iah}, \ \mathbf{V}'_{ibh} = \frac{N_{i1}}{N}\mathbf{V}_{ibh}, \ \mathbf{V}'_{ich} = \frac{N_{i1}}{N}\mathbf{V}_{ich} \qquad (3.123)$$

$$\mathbf{V}''_{iah} = \frac{N_{i2}}{N}\mathbf{V}_{iah}, \ \mathbf{V}''_{ibh} = \frac{N_{i2}}{N}\mathbf{V}_{ibh}, \ \mathbf{V}''_{ich} = \frac{N_{i2}}{N}\mathbf{V}_{ich} \qquad (3.124)$$

The transformers provide a phase shift such that the fundamental component of the voltage in zigzag winding is equal to the VSC 1 voltage both in magnitude and in phase. This condition is given by the following equations:

$$\mathbf{V}'_{ia1} - \mathbf{V}''_{ib1} = V_1\angle 0 \qquad (3.125)$$

$$\mathbf{V}'_{ib1} - \mathbf{V}''_{ic1} = V_1\angle(-120°) \qquad (3.126)$$

$$\mathbf{V}'_{ic1} - \mathbf{V}''_{ia1} = V_1\angle 120° \qquad (3.127)$$

From (3.120) to (3.127),

$$\frac{N_{i1}}{N}V_1\angle(-60°i/n) - \frac{N_{i2}}{N}V_1\angle(-60°i/n - 120°) = V_1\angle 0, \quad i = 1, 2, \ldots n-1 \qquad (3.128)$$

Equating the real and imaginary parts gives

$$\frac{N_{i1}}{N} = \cos\frac{60°i}{n} - \frac{1}{\sqrt{3}}\sin\frac{60°i}{n} \qquad (3.129)$$

$$\frac{N_{i2}}{N} = \frac{2}{\sqrt{3}}\sin\frac{60°i}{n} \qquad (3.130)$$

The zigzag windings are connected in series; the series-connected windings are connected to VSC 1 such that the $h$th-order harmonic component of the resultant voltage in phase $a$ is $\sum_{i=0}^{n-1}[(N_{i1}/N)V_h\angle(-60°ih/n) - (N_{i2}/N)V_h\angle(-60°ih/n-120°h)]$, where $N_{01} \triangleq N$ and $N_{02} \triangleq 0$. Substituting $h = 6k \pm 1$, $k = 1, 2, 3, \ldots$ gives

$$\sum_{i=0}^{n-1}\left[\frac{N_{i1}}{N}V_h\angle(-60°ih/n) - \frac{N_{i2}}{N}V_h\angle(-60°ih/n - 120°h)\right]$$

$$= \begin{cases} nV_h\angle 0 & \text{if } h = 6k \pm 1 \text{ and } k \text{ is a multiple of } n \\ 0 & \text{if } h = 6k \pm 1 \text{ and } k \text{ is not a multiple of } n \end{cases} \qquad (3.131)$$

The VSC shown in Fig. 3.32 is called $6n$-pulse VSC; for example, $n = 3$ for the 18-pulse VSC. The diagram of 18-pulse VSC is shown in Fig. 3.34.

### 3.3.3.4  Quasi Multi-Pulse VSC

The transformer requirement of a multi-pulse VSC is complicated. An alternative to multi-pulse VSC is a quasi multi-pulse VSC which consists of only wye- or

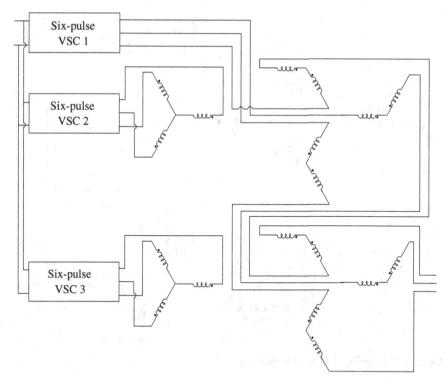

**Fig. 3.34** 18-pulse VSC

delta-connected transformers [3]. The building block of a quasi multi-pulse VSC is a 12-pulse VSC shown in Fig. 3.35. This circuit is different from the one shown in Fig. 3.31.

Equations (3.113)–(3.116) are applicable to the circuit shown in Fig. 3.35. The voltage of terminal $A'$ with respect to terminal $A'$ is

$$\mathbf{V}'_{Ah} + \mathbf{V}_{ah} = \begin{cases} 2V_h\angle 0 & \text{if } h = 1 \text{ or } h = 6k \pm 1, \ k = 2, 4, 6, \ldots \\ 0 & \text{if } h = 6k \pm 1, \ k = 1, 3, 5, \ldots \end{cases} \quad (3.132)$$

Similarly, expressions can be written for voltages of terminals $B$ and $C$ with respect to terminals $B'$ and $C'$, respectively. The order of the harmonic components in the AC-side voltages is $h = 12k \pm 1, k = 1, 2, 3, \ldots$

A quasi multi-pulse VSC is obtained by connecting 12-pulse VSCs in parallel on the DC side and in series on the AC side. The schematic diagram of a general quasi multi-pulse VSC is shown in Fig. 3.36. This VSC is called quasi $12n$-pulse VSC. For example, $n = 2$ for the quasi 24-pulse VSC.

The control signals are given such that the phase angle of the fundamental component of the AC-side voltage of the $i$th 12-pulse VSC is $-30°(i-1)/n$. Let $V_h^{(12)}$

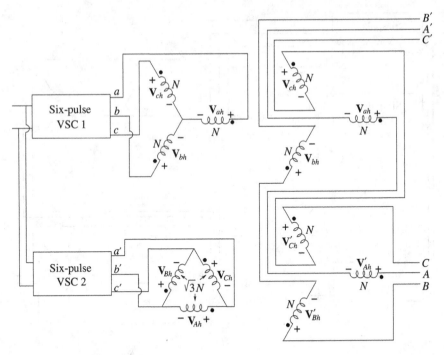

**Fig. 3.35** Building block of quasi multi-pulse VSC

**Fig. 3.36** Quasi multi-pulse
VSC

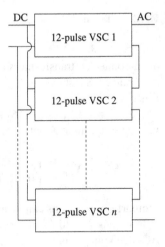

be the rms value of the voltage in each phase of a 12-pulse VSC, where $h$ is equal
to 1 for the fundamental component and $12k \pm 1, k = 1, 2, 3, \ldots$ for the harmonic
components. The rms value of the voltage in each phase of the quasi multi-pulse
VSC is

**Table 3.6** Value of $V_h/\left(nV_h^{(12)}\right)$ for different values of $n$ and $h$

|         | $h = 1$ | $h = 11$ | $h = 13$ | $h = 23$ | $h = 25$ | $h = 35$ | $h = 37$ | $h = 47$ | $h = 49$ |
|---------|---------|----------|----------|----------|----------|----------|----------|----------|----------|
| $n = 2$ | 0.9914  | 0.1305   | 0.1305   | 0.9914   | 0.9914   | 0.1305   | 0.1305   | 0.9914   | 0.9914   |
| $n = 3$ | 0.9899  | 0.1053   | 0.0952   | 0.0952   | 0.1053   | 0.9899   | 0.9899   | 0.1053   | 0.0952   |
| $n = 4$ | 0.9893  | 0.0981   | 0.0861   | 0.0648   | 0.0648   | 0.0861   | 0.0981   | 0.9893   | 0.9893   |

$$V_h = V_h^{(12)}\sqrt{\frac{2-\sqrt{3}}{2-2\cos(30°h/n)}} \qquad (3.133)$$

The value of $V_h/(nV_h^{(12)})$ for different values of $n$ and $h$ are given in Table 3.6. The AC-side voltage of a quasi multi-pulse VSC has harmonic components of all order which are present in a 12-pulse VSC. Though there is no elimination of harmonic components, the magnitudes of the harmonic components are reduced.

### 3.3.3.5 Classification of VSC

There are two types of VSC depending on whether, for a given constant DC-side voltage, the magnitude of the fundamental component of the AC-side voltages can be controlled or not [3]: type 1 VSC and type 2 VSC. The magnitude of the fundamental component of the AC-side voltages can be controlled in type 1 VSC for any given DC voltage. In a type 2 VSC, the magnitude of the fundamental component of AC-side voltages can be controlled only by changing the DC voltage. A two-level VSC is a type 2 VSC if all controllable switches are turned on and turned off only once in a cycle. A three-level VSC is an example of type 1 VSC.

## 3.3.4 STATCOM

A STATCOM is a shunt FACTS controller connected at a bus. The schematic diagram of STATCOM is shown in Fig. 3.37.

STATCOM is mainly used to regulate voltage by generating or absorbing reactive power. The reactive power is varied by varying the magnitude of the converter voltage. If losses in the converter are neglected, and the transformer does not introduce a phase shift, the fundamental component of the converter voltage and the fundamental component of STATCOM bus voltage are in phase. In a STATCOM with type 2 VSC, the converter voltage magnitude is altered by varying the DC voltage; the DC voltage is varied by drawing/supplying active power from/to the network at the STATCOM bus. Figure 3.38 shows the controller for STATCOM with type 2 VSC [3]. It consists of two control loops: the outer voltage control loop and the inner reactive current control loop. The voltage control loop is similar to the one shown in Fig. 3.18. $V$ is

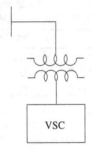

**Fig. 3.37**  Schematic diagram of STATCOM

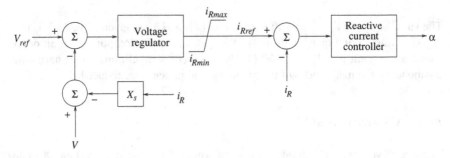

**Fig. 3.38**  Controller for STATCOM with type 2 VSC

the magnitude of the STATCOM bus voltage, and $i_R$ is the reactive current drawn by
STATCOM. $V_{ref}$ and $i_{Rref}$ are the desired values of $V$ and $i_R$, respectively. $X_s$ is
positive. In steady state, $\alpha$ is the angle by which the fundamental component of the
converter voltage leads the fundamental component of the STATCOM bus voltage;
if losses are neglected, $\alpha = 0$. In general (in steady state and during a transient),
$\alpha$ decides the instant of switching on/off of the controllable switches. Change in $\alpha$
changes the amount of active power drawn/supplied by STATCOM, thereby changing
the DC voltage. The voltage regulator and reactive current controller are typically
proportional–integral controllers. The proportional and integral gains of the voltage
regulator are negative, and the proportional and integral gains of the reactive current
controller are positive.

### 3.3.5 SSSC

A SSSC is connected in series with a transmission line. The schematic diagram of
SSSC is shown in Fig. 3.39. The three-phase winding of the transformer connected
to the VSC is wye- or delta-connected; the other three windings are connected in
series with the transmission line.

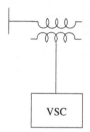

**Fig. 3.39** Schematic diagram of SSSC

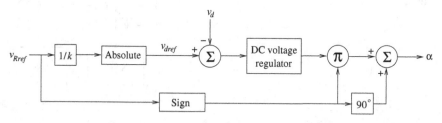

**Fig. 3.40** Controller for SSSC with type 2 VSC

SSSC generates or absorbs reactive power. The reactive power is varied by varying the magnitude of the converter voltage. If the losses in the VSC and the transformer are neglected, the fundamental component of the voltage injected by SSSC lags or leads the transmission line current by 90°. Figure 3.40 shows a part of the controller for SSSC with type 2 VSC [3]. The output of the 'Absolute' block is the absolute value of its input. The output of the 'Sign' block is 1 if its input is positive and −1 if its input is negative. The controller consists of two control loops. The outer control loop consists of a controller, for example CC controller shown in Fig. 3.22, which generates $v_{Rref}$ (instead of $X_{ref}$). $v_R$ and $v_d$ are the reactive voltage injected by SSSC and the DC-side voltage, respectively, $v_{Rref}$ and $v_{dref}$ are the desired values of $v_R$ and $v_d$, respectively. In steady state, if $v_d$ is assumed to be constant, the reactive voltage injected by SSSC is $v_R \approx k v_d$, where $k$ is a constant which includes the effect of transformer also. In steady state, $\alpha$ is the phase angle by which the fundamental component of the voltage injected by SSSC leads the fundamental component of the transmission line current; the polarity of the voltage and the direction of current are as shown in Fig. 3.40. In the inner control loop, the reactive voltage is regulated indirectly; change in $\alpha$ changes the amount of real power drawn by SSSC, thereby changing the DC voltage. The DC voltage regulator is typically a proportional–integral controller; the proportional and integral gains are positive. In steady state, if losses are neglected, $\alpha = \pm 90°$. In general (in steady state and during a transient), $\alpha$ decides the instant of switching on/off of the controllable switches.

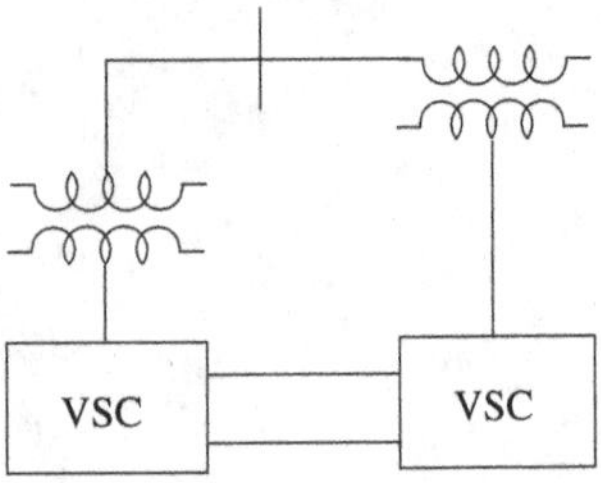

**Fig. 3.41** Schematic diagram
of UPFC

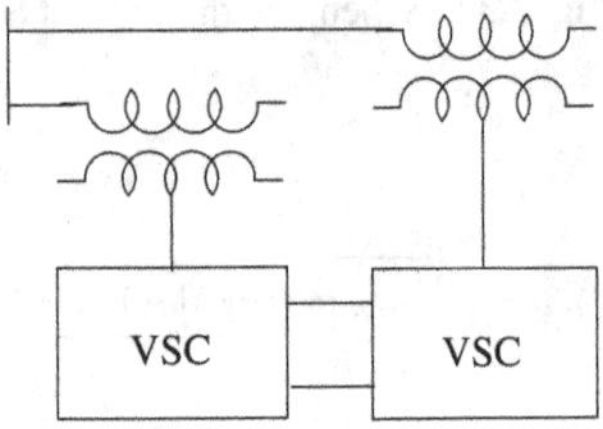

**Fig. 3.42** Schematic diagram
of IPFC

## 3.3.6 Multi-Converter FACTS Controllers

STATCOM and SSSC generate or absorb reactive power which can be controlled; in other words, these FACTS controllers have one degree of freedom. If two such VSC-based FACTS controllers are at the same location, then there can be three degrees of freedom if the two FACTS controllers are connected in parallel on the DC side, resulting in a multi-converter FACTS controller [1, 3]. The third degree of freedom is amount of power flow via the DC link. If the losses are neglected, the net active power drawn by the multi-converter FACTS controller is zero. Figures 3.41 and 3.42 show the schematic diagrams of multi-converter FACTS controllers: UPFC and IPFC. UPFC has a VSC connected in shunt through a transformer and a VSC connected in series through a transformer. IPFC has both VSCs connected in series through transformers.

## References

1. N.G. Hingorani, L. Gyugyi, *Understanding FACTS: Concepts and Technology of Flexible AC Transmission Systems* (IEEE Press, New York, 2000)
2. N. Mohan, T.M. Undeland, W.P. Robbins, *Power Electronics: Converters, Applications, and Design*, 2nd edn. (Wiley, New York, 1995)
3. K.R. Padiyar, *FACTS Controllers in Power Transmission and Distribution* (New Age International, New Delhi, 2007)
4. K.R. Padiyar, *HVDC Power Transmission Systems*, 2nd edn. (New Age International, New Delhi, 2010)

5. E.W. Kimbark, *Direct Current Transmission* (Wiley, New Delhi, 1971)
6. J. Vithayathil, *Power Electronics: Principles and Applications* (McGraw-Hill, New York, 1995)
7. P. Kundur, *Power System Stability and Control* (Tata McGraw-Hill, Noida, 1994)
8. R.M. Mathur, R.K. Varma, *Thyristor-Based FACTS Controllers for Electrical Transmission Systems* (IEEE Press, New York, 2002)
9. G. Joos, *Power Electronics: Fundamentals*, ed. by Y.H. Song, A.T. Johns Flexible AC Transmission Systems (FACTS), (IEE, London, 1999)

# Chapter 4
# Prime Movers and Excitation System

## 4.1 Prime Movers

### 4.1.1 Steam Turbine

A steam turbine has two or more turbine sections. Figure 4.1 gives the model of one type of steam turbine known as tandem compound single reheat turbine [1, 2] which has three turbine sections: high pressure (HP), intermediate pressure (IP), and low pressure (LP) turbine sections. $Y$ is valve position; $0 \leq Y \leq 1$. $T_{CH}$ is the time constant of the steam chest and the inlet piping. $T_{RH}$ is the reheater time constant and $T_{CO}$ is the time constant of the crossover piping. $F_{HP}$, $F_{IP}$, and $F_{LP}$ are the fractions of the total power generated by HP, IP, and LP turbine sections, respectively. $T_{HP}$, $T_{IP}$, and $T_{LP}$ are the torques generated by HP, IP, and LP turbine sections, respectively. $T_{max}$ is the maximum total mechanical torque.

The per unit mechanical torque generated by the turbine driving the equivalent synchronous generator with two field poles is

$$\bar{T}'_m = \frac{T_m}{T_{max}} \frac{2T_{max}}{p_f T_B} \tag{4.1}$$

### 4.1.2 Hydraulic Turbine

Figure 4.2 gives the model of the hydraulic turbine [1, 2]. $P_m$ is the mechanical power, $G$ is the gate position, and $T_w$ is the water starting time. $P_{mo}$ and $G_o$ are the initial values of $P_m$ and $G$, respectively. $G = G_o + \Delta G$ and $0 \leq G \leq 1$.

The per unit mechanical torque generated by the turbine driving the equivalent synchronous generator with two field poles is

S Krishna, *An Introduction to Modelling of Power System Components*,
SpringerBriefs in Electrical and Computer Engineering,
DOI: 10.1007/978-81-322-1847-0_4, © The Author(s) 2014

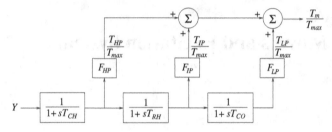

**Fig. 4.1**  Model of tandem compound single reheat turbine

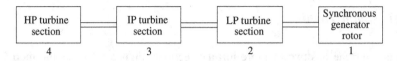

**Fig. 4.2**  Model of hydraulic turbine

| HP turbine section | IP turbine section | LP turbine section | Synchronous generator rotor |
|:---:|:---:|:---:|:---:|
| 4 | 3 | 2 | 1 |

**Fig. 4.3**  Structure of tandem compound single reheat turbine

$$\bar{T}'_m = \frac{P_{mo} + \Delta P_m}{T_B \omega} \tag{4.2}$$

## 4.2  Torsional Dynamics

A steam turbine is made up of turbine sections connected by shafts of finite stiffness. Figure 4.3 shows the structure of a synchronous generator driven by a tandem compound single reheat turbine. Each turbine section and the synchronous generator rotor can be represented by a mass. Masses 2, 3, and 4 in Fig. 4.3 represent LP, IP, and HP turbine sections, respectively, mass 1 represents the synchronous generator rotor.

Let the torques generated by LP, IP, and HP turbine sections, driving the equivalent synchronous generator with two field poles, be $T'_{LP}$, $T'_{IP}$, and $T'_{HP}$, respectively. The equations governing the torsional dynamics are [1, 2]

$$\frac{d\delta_1}{dt} = \omega_1 - \omega_o \tag{4.3}$$

$$\frac{d\delta_2}{dt} = \omega_2 - \omega_o \tag{4.4}$$

$$\frac{d\delta_3}{dt} = \omega_3 - \omega_o \tag{4.5}$$

$$\frac{d\delta_4}{dt} = \omega_4 - \omega_o \tag{4.6}$$

$$\frac{d\omega_1}{dt} = \frac{1}{J_1'} \left[ K_{12}(\delta_2 - \delta_1) - T_e' \right] \tag{4.7}$$

$$\frac{d\omega_2}{dt} = \frac{1}{J_2'} \left[ T_{LP}' + K_{23}(\delta_3 - \delta_2) - K_{12}(\delta_2 - \delta_1) \right] \tag{4.8}$$

$$\frac{d\omega_3}{dt} = \frac{1}{J_3'} \left[ T_{IP}' + K_{34}(\delta_3 - \delta_4) - K_{23}(\delta_3 - \delta_2) \right] \tag{4.9}$$

$$\frac{d\omega_4}{dt} = \frac{1}{J_4'} \left[ T_{HP}' - K_{34}(\delta_3 - \delta_4) \right] \tag{4.10}$$

where $\delta_i$ is the angular position of mass $i$ in electrical radian with respect to a reference rotating at speed $\omega_o$, $\omega_i$ is the speed of mass $i$ in electrical radian per second, $J_i'$ is the moment of inertia of mass $i$, and $K_{ij}$ is the stiffness of the shaft between masses $i$ and $j$. In some studies, the stiffness of shafts are assumed to be large and hence $\delta_1 = \delta_2 = \delta_3 = \delta_4 = \delta$ and $\omega_1 = \omega_2 = \omega_3 = \omega_4 = \omega$. Under such an assumption, (4.3)–(4.6) are equivalent to (1.132); similarly, (4.7)–(4.10) are equivalent to (1.133), where

$$J' \triangleq J_1' + J_2' + J_3' + J_4' \tag{4.11}$$
$$T_m' \triangleq T_{LP}' + T_{IP}' + T_{HP}' \tag{4.12}$$

If equations in per unit quantities are required, (4.7)–(4.10) are replaced by the following equations.

$$\frac{d\omega_1}{dt} = \frac{\omega_B}{2H_1} \left[ \bar{K}_{12}(\delta_2 - \delta_1) - \bar{T}_e' \right] \tag{4.13}$$

$$\frac{d\omega_2}{dt} = \frac{\omega_B}{2H_2} \left[ \bar{T}_{LP}' + \bar{K}_{23}(\delta_3 - \delta_2) - \bar{K}_{12}(\delta_2 - \delta_1) \right] \tag{4.14}$$

$$\frac{d\omega_3}{dt} = \frac{\omega_B}{2H_3} \left[ \bar{T}_{IP}' + \bar{K}_{34}(\delta_3 - \delta_4) - \bar{K}_{23}(\delta_3 - \delta_2) \right] \tag{4.15}$$

$$\frac{d\omega_4}{dt} = \frac{\omega_B}{2H_4} \left[ \bar{T}_{HP}' - \bar{K}_{34}(\delta_3 - \delta_4) \right] \tag{4.16}$$

where $H_i \triangleq J_i' \omega_B^2 / (2S_B)$.

## 4.3  Speed Governor

A governor is used to regulate the frequency/speed. The governor adjusts the turbine valve/gate to change the mechanical power/torque. Figure 4.4 shows a governor for the steam turbine, which is an integral controller; $K$ is positive. $\Delta Y$ is the change in

**Fig. 4.4** Isochronous governor

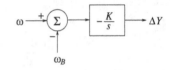

**Fig. 4.5** Governor with droop

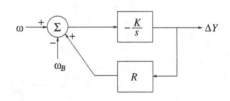

**Fig. 4.6** Governor with droop and load reference setpoint

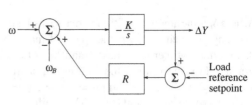

valve position. This governor is called an isochronous governor since it tries to bring the frequency back to the nominal value. The steady-state or operating speed $\omega_o$ is equal to $\omega_B$.

Isochronous governor works satisfactorily when there is a single synchronous generator in the system or when only one synchronous generator is required to respond to changes in frequency. The isochronous governor cannot be used when two or more synchronous generators in a system are required to respond to changes in frequency. This is because, due to inevitable errors in the speed measurement, which are different for different governors, a steady-state frequency would not be reached. Therefore, a feedback loop with positive gain $R$ is added as shown in Fig. 4.5 [2]; $R$ is referred to as droop. In steady state,

$$\Delta Y = -\frac{1}{R}(\omega - \omega_B) \tag{4.17}$$

If the value of $R$ of different governors are nearly equal, then for a given change in frequency, change in power output of the generators is nearly in proportion to the rating. The relation between speed and valve position can be adjusted by changing an input known as load reference setpoint which is added in the feedback loop of the governor as shown in Fig. 4.6 [2]. Therefore, in steady state,

$$\Delta Y = -\frac{1}{R}(\omega - \omega_B) + \text{Load reference setpoint} \tag{4.18}$$

With this governor, the operating speed $\omega_o$ is in general not equal to $\omega_B$.

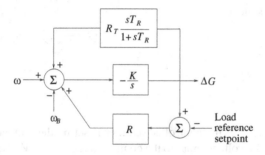

**Fig. 4.7** Governor for hydraulic turbine

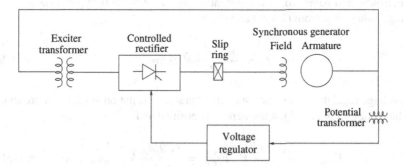

**Fig. 4.8** Schematic diagram of static excitation system

The hydraulic turbine has a peculiar response: if $\Delta G$ is a step function, from Fig. 4.2,

$$\Delta P_m = \frac{P_{mo}}{G_o}(1 - 3e^{-2t/T_w})\Delta G \tag{4.19}$$

The initial change in turbine power is opposite to that sought. Therefore, the governor for a hydraulic turbine is provided with a large temporary droop ($R_T$) with a long reset time ($T_R$), as shown in Fig. 4.7 [2].

## 4.4 Excitation System

The excitation system provides voltage to the synchronous generator field winding. This voltage is varied in order to regulate the synchronous generator terminal voltage. One type of excitation system which is prevalent nowadays is the static excitation system the schematic diagram of which is shown in Fig. 4.8 [2]. The exciter transformer steps down the voltage.

**Fig. 4.9** Model of static
excitation system

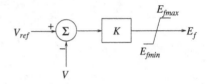

The delay angle $\alpha$ of the controlled rectifier is set by the voltage regulator. Let
the magnitude of the voltage input to the controlled rectifier be $V'$. Let it be assumed
that the field current is constant and that the input voltage to the controlled rectifier
is sinusoidal and balanced. If the instantaneous value of $E_f$ is approximated by its
average value, then from (1.194) and (3.4),

$$E_f = \frac{3\sqrt{2}M_{df}}{\pi R_f}V'\cos\alpha \tag{4.20}$$

The voltage regulator is designed such that the static excitation system is represented
by the model shown in Fig. 4.9 where $K$ is positive and

$$E_{fmax} = \frac{3\sqrt{2}M_{df}}{\pi R_f}V', E_{fmin} = \frac{3\sqrt{2}M_{df}}{\pi R_f}V'\cos\alpha_{max} \tag{4.21}$$

where $\alpha_{max}$ is the maximum value of $\alpha$. $V$ is the magnitude of the voltage at the
generator terminals and $V_{ref}$ is the desired value of $V$.

The cosinusoidal dependence of $E_f$ on $\alpha$ is negated by the appropriate choice of
$\alpha$ [3] given by

$$\alpha = \begin{cases} \cos^{-1}\dfrac{\pi R_f K(V_{ref}-V)}{3\sqrt{2}M_{df}V'} & \text{if } E_{fmin} \leq K(V_{ref}-V) \leq E_{fmax} \\ \alpha_{max} & \text{if } K(V_{ref}-V) < E_{fmin} \\ 0 & \text{if } K(V_{ref}-V) > E_{fmax} \end{cases} \tag{4.22}$$

In per unit quantities, the equation governing the excitation system is

$$\bar{E}_f = \begin{cases} \bar{K}(\bar{V}_{ref}-\bar{V}) & \text{if } \bar{E}_{fmin} \leq \bar{K}(\bar{V}_{ref}-\bar{V}) \leq \bar{E}_{fmax} \\ \bar{E}_{fmin} & \text{if } \bar{K}(\bar{V}_{ref}-\bar{V}) < \bar{E}_{fmin} \\ \bar{E}_{fmax} & \text{if } \bar{K}(\bar{V}_{ref}-\bar{V}) > \bar{E}_{fmax} \end{cases} \tag{4.23}$$

where $\bar{K} \triangleq KV_B/\psi_B$.

# References

1. K.R. Padiyar, *Power System Dynamics: Stability and Control*, 2nd edn. (BS Publications, Hyderabad, 2002)
2. P. Kundur, *Power System Stability and Control* (Tata McGraw-Hill, Noida, 1994)
3. J. Machowski, J.W. Bialek, J.R. Bumby, *Power System Dynamics: Stability and Control*, 2nd edn. (Wiley, New York, 2008)

# Erratum to: Synchronous Generator

S Krishna

## Erratum to:
## Chapter 1 in: S Krishna, *An Introduction*
## *to Modelling of Power System Components*, SpringerBriefs
## in Electrical and Computer Engineering,
## DOI 10.1007/978-81-322-1847-0_1

On page 20,

"If the total number of turns in a winding is $N$, the number of turns per field pole pair is $2N/p_f$. The inductance of the part of a winding with $2N/p_f$ turns is equal to $2/p_f$ times the expression derived above, if $\theta$ is the electrical angle. The total inductance is obtained by multiplying this expression by $p_f/2$. Hence, the inductance expressions derived above are valid even if $p_f > 2$."

Should be replaced by

"Let $N_a$, $N_f$, $N_{1d}$, $N_{1q}$, and $N_{2q}$ be the total number of turns of the equivalent sinusoidally distributed windings $a, f, 1d, 1q$, and $2q$, respectively. The number of turns, of any equivalent sinusoidally distributed winding, per field pole pair is equal

The online version of the original chapter can be found under
DOI 10.1007/978-81-322-1847-0_1

S Krishna (✉)
Department of Electrical Engineering, Indian Institute of Technology Madras, Chennai, Tamil Nadu, India
e-mail: krishnas@iitm.ac.in

S Krishna, *An Introduction to Modelling of Power System Components*,     E1
SpringerBriefs in Electrical and Computer Engineering,
DOI: 10.1007/978-81-322-1847-0_5, © The Author(s) 2014

to $2/p_f$ times the total number of turns of the respective winding. Let all the turns of any equivalent sinusoidally distributed winding be connected in series. Then, the inductance expressions derived above are valid even if $p_f > 2$, if $\theta$ is the electrical angle, and $N_a$, $N_f$, $N_{1d}$, $N_{1q}$, and $N_{2q}$ are replaced by $\sqrt{2/p_f}N_a$, $\sqrt{2/p_f}N_f$, $\sqrt{2/p_f}N_{1d}$, $\sqrt{2/p_f}N_{1q}$, and $\sqrt{2/p_f}N_{2q}$, respectively."

# Appendix A
# Solution of Linear Ordinary Differential Equations with Constant Coefficients

A set of $n$ first-order linear ordinary differential equations with constant coefficients can be written as follows:

$$\frac{dy}{dx} = Ay + u \qquad (A.1)$$

where $y$ is a $n \times 1$ vector, $A$ is a $n \times n$ matrix, $x$ is the independent variable, and $u$ is a $n \times 1$ vector dependent on $x$.

$\lambda_i$ is said to be an eigenvalue of $A$ if there exists a non zero $n \times 1$ vector $v_i$ such that

$$Av_i = \lambda_i v_i \qquad (A.2)$$

$v_i$ is called a right eigenvector of $A$ corresponding to the eigenvalue $\lambda_i$. The $n \times n$ matrix $A$ has $n$ eigenvalues $\lambda_1, \lambda_2, \ldots \lambda_n$. For a real matrix $A$, if $\lambda_i$ is a complex eigenvalue of $A$ and $v_i$ is a corresponding right eigenvector, then $\lambda_i^*$ is also an eigenvalue of $A$ and $v_i^*$ is a right eigenvector corresponding to $\lambda_i^*$.

If the eigenvalues are distinct, then the solution of (A.1) is

$$y = Te^{\Lambda(x-x_0)}T^{-1}y(x_0) + Te^{\Lambda x}\int_{x_0}^{x} e^{-\Lambda\tau}T^{-1}u(\tau)d\tau \qquad (A.3)$$

where $x_0$ is a value of $x$ for which $y$ is known, and

$$T \triangleq [v_1 \; v_2 \; \ldots \; v_n] \qquad (A.4)$$

$$e^{\Lambda x} \triangleq \begin{bmatrix} e^{\lambda_1 x} & 0 & \ldots & 0 \\ 0 & e^{\lambda_2 x} & \ldots & 0 \\ \cdot & \cdot & & \cdot \\ \cdot & \cdot & & \cdot \\ \cdot & \cdot & & \cdot \\ 0 & 0 & \ldots & e^{\lambda_n x} \end{bmatrix} \qquad (A.5)$$

S Krishna, *An Introduction to Modelling of Power System Components*,
SpringerBriefs in Electrical and Computer Engineering,
DOI: 10.1007/978-81-322-1847-0, © The Author(s) 2014

# Appendix B
# Fourier Series

The Fourier series of a periodic function $f(\omega_o t)$, with period $2\pi$, is given by

$$f(\omega_o t) = \frac{a_0}{2} + \sum_{h=1}^{\infty} [a_h \cos(h\omega_o t) + b_h \sin(h\omega_o t)] \tag{B.1}$$

where

$$a_h = \frac{1}{\pi} \int_c^{c+2\pi} f(\omega_o t) \cos(h\omega_o t) d(\omega_o t) \tag{B.2}$$

$$b_h = \frac{1}{\pi} \int_c^{c+2\pi} f(\omega_o t) \sin(h\omega_o t) d(\omega_o t) \tag{B.3}$$

$c$ can be chosen arbitrarily. $a_0/2$ is the average value of $f(\omega_o t)$. The rms value of the $h$th-order harmonic component of $f(\omega_o t)$ is $\sqrt{(a_h^2 + b_h^2)/2}$. The Fourier series of $f(\omega_o t)$ is also given by

$$f(\omega_o t) = \sum_{h=-\infty}^{\infty} c_h e^{jh\omega_o t} \tag{B.4}$$

where

$$c_h = \frac{1}{2\pi} \int_c^{c+2\pi} f(\omega_o t) e^{-jh\omega_o t} d(\omega_o t) \tag{B.5}$$

From (B.2), (B.3), and (B.5),

$$c_0 = \frac{a_0}{2} \tag{B.6}$$

$$c_h = \frac{a_h - jb_h}{2}, \quad h = 1, 2, 3, \ldots \tag{B.7}$$

S Krishna, *An Introduction to Modelling of Power System Components*,
SpringerBriefs in Electrical and Computer Engineering,
DOI: 10.1007/978-81-322-1847-0, © The Author(s) 2014

$$c_h = \frac{a_{-h} + jb_{-h}}{2}, \quad h = -1, -2, -3, \ldots \tag{B.8}$$

The rms value of the $h$th-order harmonic component of $f(\omega_o t)$ is $\sqrt{2}|c_h|$.

If $f(\omega_o t)$ is an odd function, i.e., $f(-\omega_o t) = -f(\omega_o t)$,

$$a_h = 0 \tag{B.9}$$

$$b_h = \frac{2}{\pi} \int_0^\pi f(\omega_o t) \sin(h\omega_o t) d(\omega_o t) \tag{B.10}$$

If $f(\omega_o t)$ is an even function, i.e., $f(-\omega_o t) = f(\omega_o t)$,

$$a_h = \frac{2}{\pi} \int_0^\pi f(\omega_o t) \cos(h\omega_o t) d(\omega_o t) \tag{B.11}$$

$$b_h = 0 \tag{B.12}$$

If $f(\omega_o t)$ possesses half-wave symmetry, i.e., $f(\omega_o t + \pi) = -f(\omega_o t)$,

$$a_h = \begin{cases} 0 & \text{if } h = 0, 2, 4, \ldots \\ \dfrac{2}{\pi} \displaystyle\int_c^{c+\pi} f(\omega_o t) \cos(h\omega_o t) d(\omega_o t) & \text{if } h = 1, 3, 5, \ldots \end{cases} \tag{B.13}$$

$$b_h = \begin{cases} 0 & \text{if } h = 2, 4, 6, \ldots \\ \dfrac{2}{\pi} \displaystyle\int_c^{c+\pi} f(\omega_o t) \sin(h\omega_o t) d(\omega_o t) & \text{if } h = 1, 3, 5, \ldots \end{cases} \tag{B.14}$$

$f(\omega_o t)$ is said to possess quarter-wave symmetry if $f(\omega_o t)$ possesses half-wave symmetry and there exists a $\phi$ such that $f(\omega_o t + \phi) = f(-\omega_o t + \phi)$. If $f(\omega_o t)$ possesses quarter-wave symmetry, then for the function $f(\omega_o t + \phi)$,

$$a_h = \begin{cases} 0 & \text{if } h = 0, 2, 4, \ldots \\ \dfrac{4}{\pi} \displaystyle\int_0^{\pi/2} f(\omega_o t + \phi) \cos(h\omega_o t) d(\omega_o t) & \text{if } h = 1, 3, 5, \ldots \end{cases} \tag{B.15}$$

$$b_h = 0 \tag{B.16}$$

It is to be noted that the rms values of a harmonic component of $f(\omega_o t)$ and $f(\omega_o t + \phi)$ are equal.

The harmonic components of order 3 and its multiples are called triplen harmonic components. The triplen harmonic components of $f(\omega_o t)$, $f(\omega_o t - 2\pi/3)$, and $f(\omega_o t + 2\pi/3)$ are equal. Hence, the triplen harmonic component of $f(\omega_o t)$ is equal to one third of the triplen harmonic component of $f(\omega_o t) + f(\omega_o t - 2\pi/3) + f(\omega_o t + 2\pi/3)$. Therefore, the triplen harmonic components are equal to zero if $f(\omega_o t) + f(\omega_o t - 2\pi/3) + f(\omega_o t + 2\pi/3) = 0$.

# About the Author

S Krishna is Assistant Professor of Electrical Engineering at Indian Institute of Technology Madras, Chennai, India. He received B.E. degree from Bangalore University in 1995 and M.E. and Ph.D. degrees from Indian Institute of Science, Bangalore, in 1999 and 2003, respectively. He worked with Kirloskar Electric, Bangalore, from 1995 to 1997 and M.S. Ramaiah Institute of Technology, Bangalore, from 2003 to 2008. His areas of research interest are power system dynamics and control. He is an Associate Editor of SADHANA, a Journal of the Indian Academy of Sciences.

# About the Book

The brief provides a quick introduction to the dynamic modelling of power system components. It gives a rigorous derivation of the model of different components of the power system such as synchronous generator, transformer, transmission line, FACTS, DC transmission system, excitation system, and speed governor. Models of load and prime movers are also discussed. The brief can be used as a reference for researchers working in the areas of power system dynamics, stability analysis, and design of stability controllers. It can also serve as a text for a short course on power system modelling or as a supplement for a senior undergraduate/graduate course on power system stability.

S Krishna, *An Introduction to Modelling of Power System Components*,
SpringerBriefs in Electrical and Computer Engineering,
DOI: 10.1007/978-81-322-1847-0, © The Author(s) 2014

# Index